T0130811

COMMISSIONERS OF THE FCC

1927–1994

Edited by
Gerald V. Flannery
University of Southwestern Louisiana
Lafayette, La.

UNIVERSITY
PRESS OF
AMERICA

Lanham • New York • London

Copyright © 1995 by
University Press of America® Inc.
4720 Boston Way
Lanham, Maryland 20706

3 Henrietta Street
London WC2E 8LU England

All rights reserved
Printed in the United States of America
British Cataloging in Publication Information Available

Library of Congress Cataloging-in-Publication Data

Commissioners of the FCC, 1927–1994 / edited by Gerald V.
Flannery.
p.   cm.
1. United States. Federal Communications Commission—Biography.
2. Telecommunication policy—United States—History.   I.
Flannery, Gerald V.
HE7781.C63   1994   384'.092'273—dc20   94–5408 CIP
[B]

ISBN 0–8191–9669–X (cloth : alk. paper)

 The paper used in this publication meets the minimum requirements of
American National Standard for Information Sciences—Permanence
of Paper for Printed Library Materials, ANSI Z39.48–1984.

This book is dedicated to

my loving wife

Laura

whose support is

constant.

A special word of thanks to those graduate students who helped with this project.

Michael Konczal	Marisol Ochoa Konczal
Bobbie DeCuir	James Nunez
David A. Male	Peggy Voorhies
Richard E. Robinson	Ricky L. Jobe
Carla P. Coffman	Ken Ditto
Mary Syrett	Fran DeLaun
Kim Bourque	Brenda Trahan
Robyn Wimberly Blackwell	Richard Tanory
Kelli M. Vallot	Andrea LeBlanc
Joe Lynch	Brian P. Atkinson

And good friend Edwin A. Meek, Ph.D., University of
      Mississippi.

Special thanks also go to:

    1) The Public Information Office of the Federal
        Communications Commission.

    2) Broadcasting magazine's Chronology of Events
        published in 1971 and 1993 and its
        excellent coverage of the Commission's actions
        over the years.

TABLE OF CONTENTS

FEDERAL RADIO COMMISSIONERS   vii

Sykes, Eugene O.	(1927-1934)	Mississippi	1
Bellows, Henry A.	(1927-1927)	Minnesota	4
Brown, Thad H.	(1932-1934)	Ohio	7
Bullard, Admiral W.H.C.	(1927-1927)	Pennsylvania	10
Caldwell, Orestes H.	(1927-1929)	New York	13
Dillon, Col John F.	(1927-1927)	California	16
Hanley, James H.	(1933-1934)	Nebraska	19
Lafount, Harold A.	(1927-1934)	Utah	23
Pickard, Sam	(1927-1929)	Kansas	25
Robinson, Ira E	(1928-1932)	West Virginia	28
Saltzman, Gen. C. Mck.	(1929-1932)	Iowa	31

FEDERAL COMMUNICATIONS COMMISSIONERS

Gary, Hampson	(1934-1934)	Texas	33
Stewart, Irvin	(1934-1937)	Texas	36
Payne, George H.	(1934-1943)	New York	39
Case, Norman S.	(1934-1945)	Rhode Island	42
Walker, Paul Atlee	(1934-1953)	Oklahoma	45
Prall, Anning S.	(1935-1937)	New York	48
Craven, T.A.	(1937-1944, 1956-1963)	D.C.	51
McNinch, Frank R.	(1937-1939)	North Carolina	55
Thompson, Frederick I.	(1939-1941)	Alabama	58
Fly, James L.	(1939-1944)	Texas	60
Wakefield, Ray C.	(1941-1947)	California	63
Durr, Clifford J.	(1941-1948)	Alabama	66
Jett, Ewell K.	(1944-1947)	Maryland	69
Porter, Paul A.	(1944-1946)	Kentucky	72
Hyde, Rosel H.	(1946-1969)	Idaho	75
Denny, Charles R.	(1945-1947)	D.C.	78
Wills, William H.	(1945-1946)	Vermont	81
Webster, Edward M.	(1947-1956)	D.C.	84
Jones, Robert F.	(1947-1952)	Ohio	87
Coy, Albert W.	(1947-1952)	Indiana	90
Sterling, George E.	(1948-1954)	Maine	93
Hennock, Frieda B.	(1948-1955)	New York	96
Bartley, Robert T.	(1952-1972)	Texas	99

Merrill, Eugene H.	(1952-1953)	Utah	102
Doerfer, John C.	(1953-1960)	Wisconsin	105
Lee, Robert E.	(1968-1973	Illinois	108
McConnaughey, George	(1954-1957)	Ohio	111
Mack, Richard A.	(1955-1958)	Florida	114
Ford, Frederick W.	(1957-1964)	West Virginia	117
Cross, John S.	(1958-1962)	Arkansas	120
King, Charles H.	(1960-1961)	Michigan	123
Minow, Newton N.	(1961-1963)	Illinois	126
Henry, E. William	(1962-1966)	Tennessee	129
Cox, Kenneth A.	(1963-1970	Washington	132
Leovinger, Lee	(1963-1968)	Minnesota	135
Wadsworth, James J.	(1965-1969)	New York	138
Houser, Thomas J.	(1971-1971)	Illinois	141
Wells, Robert	(1969-1971)	Kansas	144
Johnson, Nicholas	(1966-1973)	Iowa	147
Lee, H. Rex	(1968-1973)	D.C.	151
Burch, Dean	(1969-1974)	Arizona	154
Reid, Charlotte T.	(1971-1976)	Illinois	157
Wiley, Richard E.	(1972-1977)	Illinois	160
Hooks, Benjamin L.	(1972-1977)	Tennessee	163
Quello, James H.	(1974-	Michigan	166
Robinson, Glenn O.	(1971-1976)	Minnesota	169
Washburn, Abbott M.	(1974-1982)	Minnesota	172
White, Margita E.	(1976-1979)	Virginia	175
Ferris, Charles D.	(1977-1981)	Massachusetts	178
Brown, Tyrone	(1977-1981)	D.C.	181
Fogarty, Joseph R.	(1976-1983)	Rhode Island	184
Jones, Anne	(1979-1983)	Massachusetts	187
Sharp, Stephen A.	(1982-1983)	Ohio	191
Rivera, Henry M.	(1981-1985)	New Mexico	194
Fowler, Mark S.	(1981-1987)	Canada / US	197
Dawson, Mimi W.	(1981-1987)	Missouri	200
Patrick, Dennis R.	(1983-1992)	California	202
Dennis, Patricia D.	(1986-1989)	New Mexico	205
Sikes, Alfred C.	(1989-1993)	Missouri	208
Marshall, Sherrie. P.	(1989-1993)	Florida	211
Barrett, Andrew C.	(1989-	Illinois	214
Duggan, Ervin S.	(1990-	Illinois	218
Hundt, Reed	(1993-	Michigan	221
Chong, Rachelle	(1994-	California	223
Ness, Susan	(1994-	D.C.	225

Preface

The Federal Radio Commission and the subsequent Federal Communications Commission were two government agencies charged with regulating interstate and foreign communication for America.

The government has been involved in broadcasting since its early days when the tests of wired radio and wireless were under way, first because of its interest in ship to ship and ship to shore communication, with the Radio Act of 1912, for navigation and safety purposes, then more heavily in World War I. Communication in wartime is always a critical aspect of battle, strategy and planning and the government poured money and time into developing anything that would give its troops an edge in warfare.

The early days of commercial radio, after World War I, were chaotic as too many licenses were issued in the twenties, there were too many stations, a lot of drifting off assigned frequencies, patent fights, equipment with poor technical quality, transmitters being moved at will to new locations, and struggles over advertising, program content and music royalties, to name a few.

The industry was growing at all levels but could not agree on how to operate, much less cooperate. Finally, in desperation, industry leaders got the government to call a series of radio conferences, to try to deal with the mounting problems. They were unsuccessful in conferences one, two and three,which managed to flesh out the problems but not solve them; however, they did manage to reach some agreement by the end of conference number four. What emerged was the Radio Act of 1927 setting up the Federal Radio Commission.

The purpose of the regulation was to provide for the orderly development and operation of broadcasting services, to make available a rapid, efficient, nation-wide and world-wide telegraph and telephone service at reasonable charges, to promote the safety of life and property through the use of wire and radio communication; and to employ communication facilities for strengthening the national defense. The FRC and the FCC were independent Federal establishments, created by Congress, and which reported directly to Congress. Its rules and regulations had the force of law, since it was a federal agency, until those rules were tested in court and approved or rejected.

Early jurisdiction of wire and radio communication was shared at first by the Department of Commerce, the Post Office Department , the Interstate Commerce Commission and, later, the Federal Radio Commission. Rapid growth in all areas of the communication field necessitated better coordination of those regulatory functions, so, on June 19, 1934, the Federal Communications Commission was created, governing not only the United States but Guam, Puerto Rico and the Virgin Islands.

The FCC allocated bands of frequencies to non-government radio and television services, assigning frequencies to individual stations; licensing and regulating radio and television stations and station operators; regulating common carriers engaged in interstate and foreign communication by telegraph and telephone; promoting safety through the use of those mediums on land, water and in the air; encouraging more effective and widespread use of radio and television; harnessing radio and television service to the national defense; and helping to establish and support communications systems in space.

The five Commissioners who run the agency (once there were seven) are appointed by the President and have to be approved by the Senate. No Commissioner can have any financial interest in any of the businesses that the FCC regulates. The normal appointment was seven years (later changed to five) except if someone is appointed to fill out an unexpired term. The President appoints one of the Commissioners as Chairman and he / she serves at the pleasure of the President, but policy decisions are made by the Commission as a whole.

The FCC is organized into four bureaus: 1)Field Engineering; 2) Common Carrier; 3) Safety and Special Radio Services; and 4) Broadcast.; and into eight offices: a) Executive Director; b) Secretary; c) Chief Engineer; d) General Counsel; e) Review Board; f) Hearing Examiners; g) Opinions and Review; and h) Reports and Information.

The FCC itself is responsible for numerous things: 1) assigning frequencies, station power, call letters and time of operation for stations and operators; 2) considering license applications; 3) inspecting communications equipment; 4) approving transfers of ownership; and 5) renewing licenses.

The FCC, through the Commission and its offices, is involved in both national and international matters concerning communications. It licenses radiotelephone and radiotelegraph circuits, telephone and telegraph facilities, supervises charges, practices, classifications, of radio, telephone, wire and cable, television, cablevision and satellites, including space communication, and the mergers of any and all of those entitles. FCC rules and regulations have the force of law unless challenged in court.

The field staff of the FCC monitors the spectrum to see that station operation meets the technical requirements; it inspects stations of all types; it conducts operator
exams and issues permits to those who pass; it locates and shuts down illegal or unauthorized transmitters; it traces and halts any interference with the airwaves; and it obtains and analyzes technical data for the Commission.

It is charged to see that the communication(s) facilities in the United States operate in "the public interest, convenience and necessity." It does this by all of the foregoing and in ways that affect content without controlling it: things like the Fairness Doctrine (now rescinded), the Equal Time rule, Section 415, the Indecency clause, cross-ownership of media properties, concentration of ownership of properties, quantity and type of ownership of communication(s) facilities, and much more.

The Federal Communications Commission has been a source of controversy since its inception, often drawing the fire of broadcasters, sometimes the praise, sometimes moving too slowly to regulate, or guide, or shape an industry that is exploding in all directions, with technology creating new channels/ types/ levels of communication faster than existing regulations can handle, or new ones can be developed.

The post of Commissioner has been viewed as a mixed blessing by many FCC members who find the job challenging, interesting, frustrating, time-consuming, difficult and rewarding. This book is about the people who have accepted that responsibility.

By Gerald V. Flannery, Ph.D. based on FCC documents.

SYKES, EUGENE O.
1927-1934

Eugene Octave Sykes, a soft spoken Mississippian, brought promise out of the chaos of early radio and set the stage for the regulation of broadcasting in the public interest. Sykes was one of the original members of the Federal Radio Commission and he eventually became its Chairman. He later became the first Chairman of the Federal Communications Commission. Sykes, more than anyone else is given credit for straightening out the early tangles of radio.

Sykes was born in Aberdeen, Mississippi on July 16, 1867, the son of Eugene Octavious Sykes, a judge, legislator, former Captain in the Confederate army, and member of the Constitutional Congress of 1890. Sykes attended the public school of Aberdeen and studied at Bell Buckle College in Tennessee, and at St. John's College, a noted institution of learning and the third oldest college in the United States. Add to this a stint at the U.S. Naval Academy at Annapolis and some time at the University of Mississippi. He graduated from "Ole Miss" in 1897 and began a private law practice in Aberdeen. He then went on to become a Democratic Presidential Elector at Large for the state of Mississippi in 1904, and in 1916, the infamous governor, Theodore G. Bilbo, appointed him to the Mississippi Supreme Court. Sykes also served on the State Democratic Executive Committee. He voluntarily retired from the bench in 1924 and resumed his law practice. President Calvin Coolidge appointed him to the newly formed Federal Radio Commission in March, 1927.

Radio had grown unchecked in the decade prior to 1927. The industry was riddled with offensive advertising, wave piracy, monopolies, the creation of networks, the sale of a station's license, the use of unpaid performers and royalty-free material. Coolidge called on Sykes to "lay down the law" and the small, slim, unusually quiet man was thrown into the maelstrom without any real experience in radio. Sykes was appointed to a second term on the Commission by Herbert Hoover on February 24, 1930, then at 68 years of age, became the first Chairman of the FCC in 1934. A New York Times newspaper article years later

(1945) wrote of the challenges the FRC faced saying that it issued more than 35,000 licenses from 1927 to 1933, with 608 of those being for radio stations, and that the former judge was invaluable in untying the legal knots.

The Official and Statistical Register of Mississippi referred to him as a poised, deliberate, sympathetic and courageous person, someone who always held the interests of the listener above all else, who stood for the rights of independent broadcasting and who defended America's rights in the international scramble for wave lengths.

Sykes, on March 17, 1927, speaking over a 29-station hook-up, only a short time after his appointment, told the radio audience that the most significant feature of the new radio law was the dominant influence of the "public interest." Sykes outlined a premise that would be interpreted and reinterpreted, in the decades to follow, by Congress, the Federal Communications Commission, broadcasters and the courts. The New York Times for March 18, 1927 said: "Judge Sykes declared that neither he nor any of his associates was under any obligation, political or otherwise, except to the radio public...and this (idea) is the constitutional basis for every action the Commissioner will take."

The Times article went on to summarize Sykes: "Our hope is to interfere...just as little as we can...which means the broadcasters...must help us. We believe they will recognize...they cannot have everything they want...and that they will achieve everything they want by serving a satisfied public."

Judge Sykes, as Vice Chairman of the FRC in 1929, testified in support of congressional efforts which finally established the Federal Communications Commission.

Broadcasting magazine, first published in October 1931, told how the FRC had granted full power (50KW) to nine stations, bringing to 23 the number of full power stations. On June 15, 1932 the FRC reported to Congress that "to eliminate the use of radio facilities for commercial advertising purposes, will, if adopted, destroy the present system of broadcasting." When the FCC Act became law in 1934, the 68-year-old former jurist, appointed earlier by President Franklin D. Roosevelt, was elected its first Chairman, Sykes served as Chairman for only eight months, when he stepped down to become chair of the FCC's radio division.

In 1936, Sykes called on the National Association of Broadcasters to improve broadcast services to rural America. The following year he addressed the NAB'S national convention. The New York Times, in its June 27, 1937 issue quoted him:

The Commission, having determined that you are a proper party to serve the public in your community and having granted you this gracious privilege in the form of a license, this duty then devolves upon you to render the best public service that you can. You assume, by the acceptance of this license, a great responsibility of public service.

The Commission had to organize itself and the broadcasting and telecommunications industry had to be brought under regulatory control. The FCC issued orders requiring licensees to file information regarding ownership of stations, investigated and subsequently lowered long distance telephone charges, established new engineering standards for AM stations, completed hearings on radio frequency allocations and negotiated with other North American countries regarding the use of the radio spectrum in an effort to avoid interference across national boundaries."

Sykes played a major role in these developments, particularly with domestic and international radio. President Coolidge sent him to represent the United states at the North American Radio Conference held in Ottawa in 1929 and President Hoover appointed him chairman of the American delegation to the International Radio Conference in Madrid in 1932. The Madrid meeting, which lasted four months, added two other accomplishments: 1) he was elected chairman of the important technical committee; and 2) he persuaded the delegates to institute English, along with French, as the official language of the conference.

Judge Sykes resigned from the FCC on April 6, 1939, and became a partner in a private law practice in Washington. There followed a number of tributes from people in politics and the industry. For example, Eliot C. Lovett, president of the FCC Bar Association, called him the Association's most illustrious member; Rosel H. Hyde, FCC general counsel, said Sykes exemplified dignity, ability, respect, and confidence; and NAB president J. Harold Ryan praised him as a man who understood the needs of broadcasting and public service. Sykes died of a heart attack on June 21, 1945.

Portions of this article appeared in FEEDBACK, the Journal of the Broadcast Education Association.

By Gerald V. Flannery, Ph.D. and Edwin A. Meek, Ph.D.

BELLOWS, HENRY
1927-1928

Henry Bellows came from an educated, liberal, and intellectual background. He had a Ph.D. from Harvard (1910), was the author of four books, served as editor, and also as a translator. He felt radio needed to encompass public service and worked for that idea all through his short term. He came to the Federal Radio Commission as CBS went on the air with a network of 16 stations.

Bellows was born in September 1885, the son of John and Isabelle Bellows of Portland, Maine. He married Mary Sanger in 1911 and had two children, Eleanor and Charles. Following his wife's death, Bellows married Alice Eels in April 1936.

On April 29, 1927, in a speech before the League of Women Voters, Bellows explained that the lack of censorship in Radio Law compelled the listener to govern programming, since government was forbidden to do it. In this regard, he felt that the audience must demand not only first rate entertainment, but public service information as well. Should the audience or the radio industry fail to meet this challenge, Bellows foresaw more government regulation, something he felt was a deplorable alternative.

Bellows was one of the first Commissioners appointed in 1927 to the newly formed Federal Radio Commission (predecessor to the Federal Communications Commission). In its infancy, the organization's guidelines were born out of a bill authorized by Washington Democratic Senator Dill.Certainly, Bellows' credentials helped him in his new post: his work in radio and newspapers, as manager of WCCO Radio (1925), still one of the leading stations in the country, or as managing editor of "The Bellman" (1912 -1919) and "The Northwestern Miller" (1914-1925), and as Music Critic for "The Minneapolis Daily News" (1921-1923), all qualified him as an expert in the eyes of President Calvin Coolidge.

Those early days of the FRC were filled with controversy since many did not want the government involved in broadcasting at all, while others thought it ought to control the content of programs, not just the technical aspects. The early twenties saw a series of radio conferences, four in all, among interested parties, brought together by the government, to solve the myriad problems as radio developed. There were patent fights, frequency fights, and interference with other stations, planned and accidental, as licensees changed locations and wandered off assigned frequencies because of the poor technical quality of the early equipment. There was also the problem of power: the first radio sets were battery-powered, using a wet cell system, that was messy and frequently leaked. Finally, in the late twenties, residential America became electrified and the listener finally had a reliable source of power for a radio. This was the milleau Bellows stepped into as a virgin commissioner.

In October, only eighteen months after his appointment to a three year term, Bellows left the Commission, but not the radio industry. A month before he left, Bellows addressed the National Association of Broadcasters, on September 21, 1927, one of his last speeches as an FRC Commissioner. Bellows was harsh in his criticism of radio programming emphasizing that only a "few stations had any clear conception of what constituted public convenience, necessity and interest."

During his brief tenure on the FRC, it faced the fact that many things about the radio business were already accepted practices, put in place before the Commission got started. Things like the network system of radio with NBC and CBS. In fact, one month after he joined the FRC, CBS went on the air with a basic network of sixteen stations. Add to this the fact that a station license had been sold (WEAF) to someone else for one million dollars, and then there was the acceptance and use of advertising on radio. These practices became "grandfathered" in.

Bellows felt each station needed an excellent program producer and called it a necessity, in his opinion, to meet public demands regardless of technology. Bellows stressed individuality in programming, an objective obtainable only through a study of each station's audience. This included analysis of the listener's public service needs. Radio, newspapers and magazines were the only means of mass communication, in those days, so Bellows' stance was really ahead of its time. Perhaps it was this attitude which led Bellows back into the private sector prematurely.

In 1927, he served as Technical Advisor to the International Radio Telegraph Conference and, after he left the FRC, was a Director of the National Association of Broadcasters from 1928-1935. Bellows managed Northwestern Broadcasting, Inc. (1929-1934) and for four years was a Vice President of Columbia Broadcasting System (1930-1934). He extended his influence across continents initiating a transatlantic exchange agreement for radio programs (1930). In 1936, he became Director of Public Relations for General Mills, Inc.

Bellows, a scholar, was also known for his translation of the *Poetic Edda* for the American-Scandinavian Foundation and the *Historia Calamitatum* of Abeland. He wrote four books and also taught as an Assistant Professor of English at Harvard from 1906-1909 and then taught Rhetoric at the University of Minnesota from 1910-1912.

Unfortunately Bellows scholastic, public, and commercial goals were ended by an early death at fifty-four years of age in December 1939.

By Gerald V. Flannery, Ph.D. and Peggy Voorhies, M.S. candidate.

BROWN, THAD H.
1932-1934

Confrontation and change marked the years that Thad Brown was involved in the government regulation of broadcasting. During the years 1932,1933, and 1934 his nomination and renomination was challenged each year until the Federal Radio Commission gave way to the Federal Communications Commission. Brown was born in Lincoln Township, Morrow County, Ohio, on January 10, 1887. He was raised on a farm, the son of William Henry and Ella Brown. As a young man he attended a country school and, after graduating from high school, Brown taught in the village school for one year. He then went on to further his own education by attending, and earning degrees at Ohio Wesleyan University and Ohio State University. From 1909 to 1911 Brown served as Journal Clerk of the Ohio House of Representatives. His interest in law grew and he was admitted to practice law in Ohio in June 1912 after receiving a law degree from Lincoln University.

Brown married Marie Thrailkill in 1915. They has one son, Thad, Junior. The first World War called him away from his work as a lawyer from July 1917 to February 1919. He served in the infantry for 19 months, entering as a Captain and retiring as a Major. Brown continued his military service as a Colonel in the Officers Reserve Corps.

In the early to mid 1920's Brown became very active in politics. He was appointed by Governor James M. Cox as a member of the State Civil Service Commission of Ohio on February 1,1920. He served on that commission until resigning to qualify for the post of Secretary of State of Ohio. From 1920 to 1921 he served as Chairman of the Americanization Committee of the American Legion of Ohio. His stint as Secretary of State of Ohio ran from January 8, 1923 to January 10, 1927. He ran for Governor of Ohio in 1926 but was defeated by a few thousand votes.

Brown became active in broadcasting in April 1927 when he became president of Cleveland Radio Broadcasting Corporation and manager of radio station WJAY. He left that position in February of 1928 and during that year he was a delegate to the Republican National Conference and a Presidential Elector. The following year, on September 12, he became chief counsel of the Federal Power Commission.

Brown combined his experience in law and broadcasting, in December 1929, when he became General Counsel of the Federal Radio Commission. He kept that position until 1932 when he resigned so that President Hoover might name him to the FRC. That, and the fact that he had worked to elect Hoover in 1928, drew fire in the January 22, 1932 issue of The New York Times:

> Hardly had Mr. Hoover's choice of Thad Brown to be a member of the Radio Commission reached the Senate before Chairman Couzens of the Interstate Commerce Committee announced he would contest the appointment, which goes to his committee.

Couzens claimed that the appointment was a political one and charged the former Ohio Secretary with malfeasance while in that office. The debate continued, until finally, ten weeks later, Colonel Thad Brown was sworn into office by Judge Charles S. Hatfield of the Court of Customs and Patent Appeals, one of Brown' closest friends.

Some of the major issues facing the FRC during Brown's tenure were items like the synchronization of radio signals, cooperation among members of the telephone, telegraph and broadcast industries, and the changeover from the FRC to the FCC.

Broadcasting magazine, in its 1971 chronology of radio events, listed some of the items the FRC dealt with during his tenure. They were: the continued development by Dr. V.K. Zworkin of his kinescope or cathode-ray television receiver; the struggle between ASCAP and station owners represented by the NAB; and that phonograph companies were labelling their records "Not For Use On Radio." Two out of every five American households had a radio, despite the depression; radio placed all its facilities at the disposal of President Franklin Roosevelt during the banking crisis; and Associated Press members voted to ban network broadcasts of AP news.

The early 1930's was a period when America was in the grip of a nationwide economic depression, one that saw hundreds of thousands of people out of work and without any government safety net to provide for the jobless and the homeless. The stock market crash was still being felt and banks were closing in record numbers. Still, in the midst of all this radio grew and prospered. President Franklin D., Roosevelt was using radio to speak to America and explain his recovery program. The networks were developing new program formats for radio and the FCC was waiting in the wings.

Brown left the FCC amidst the same kind of political fireworks that greeted his entrance. President Roosevelt had reappointed Brown to the FCC but the move aroused considerable opposition and was finally blocked by a Senate committee. Brown resigned the FCC in 1934, its maiden year.

Brown, a 32nd degree Mason, was a joiner belonging to many organizations; the American Bar Association, the Ohio State Bar Association and the Federal Bar Association; the American Academy of Political and Social Sciences; the international Committee on Radio, Executive Committee, American Section; the American Legion of Ohio; the Presbyterian Church, and the Shriners. He died on February 28, 1941.

By Gerald V. Flannery, Ph.D. and Richard E. Robinson, M.S.

BULLARD, WILLIAM H. G.
1927-1927

W. H. C. Bullard has been referred to as the "father of radio." He was instrumental in the early development of radio and was one of the original members of the Federal Radio Commission.

Born in Media, Pennsylvania, on December 5, 1866, Bullard went on to serve in the U. S. Navy for 36 years. He served in the American Division with the British Grand Fleet during World War I. Bullard was very concerned with the protection and development of radio as applied to marine service. He warned all coastal broadcasters not to interfere with marine distress calls. If stations were found to be interfering, their frequencies would be changed. Bullard argued strongly for the protection of the rights of amateur radio operators.

Bullard was influential in allowing press messages to be transmitted over Navy transmitters. This made news from Hawaii and the Philippines accessible to the U.S.

Guglielmo Marconi's patent for wireless telegraphy in 1901 set the stage for the wireless revolution. In 1897, the British Marconi Company was formed to acquire the title to Marconi's patents. A subsidiary, American Marconi was formed in the United States in 1899 and soon gained control of nearly all of commercial wireless communications. Before World War I, General Electric, Westinghouse, and the Western Electric Company (a subsidiary of AT&T) won several radio patents. A patent war riddled the radio industry until the government took over all wireless stations during World War I. The government had industry pool their inventions for the war effort in return for assurance that the government would offer legal protection against patent suits. After the war, in 1919, British Marconi began negotiating with General Electric for exclusive rights to the Alexanderson alternator. The deal was nearly complete when Bullard stepped into the proceedings.

As the Director of Naval Communication (1919), he protected the patent of the Alexanderson alternator made by General Electric. He assured that the patent would be retained by Americans and not fall into foreign control. The alternator was the most reliable device available for sending radio communications to ships and across the ocean. Bullard urged the Secretary of the Navy to retain the patent and to form a new company to enter the communications field. The end result was the organization of the Radio Corporation of America, later the parent of NBC, the National Broadcasting System. President Woodrow Wilson detached Bullard from his naval responsibilities in order to sit on the Board of Directors.

Bullard retired from active service with the Navy in 1922. In 1927, President Calvin Coolidge picked Bullard to head the FRC. He was one of the original five board members; the others were Bellows, Caldwell, Dillon, and Sykes.

The industry was clamoring for some kind of control but had failed several times at self regulation. There were disagreements over patents, hours of operations, transmitter sites, the quality of the equipment, how much control the government should have, how to pay for radio, and who would be paid for the music and songs played on the air. A series of radio conferences in the early twenties only led industry leaders to the conclusion that the government must get more involved than just in granting licenses. Bullard witnessed the formation of AT&T, NBC, CBS, and during his FRC tenure, the organization of the Columbia Phonograph Broadcasting System in April 1927. The CPBS functioned as the sales agency of United. In addition, Bullard's tenure was marked by President Herbert Hoover's unsuccessful efforts to regulate radio. From July 1926 to February 1927, nearly two hundred new radio stations began broadcasting without frequency or power level restrictions.

In response to the uncontrollable radio market, the Radio Act of 1927 was initiated in February of 1927. The Act proclaimed that the airwaves belonged to the people and were to be used by individuals only with the authority of short-term licenses granted by the government. The Act created a mad scramble for license applications by stations who wished to operate under the new law. The Federal Radio Commission was formed to administer the Radio Act.

By 1929, radio programs, such as "Amos 'n' Andy" and "The Goldbergs," began broadcasting. It was a period when new radio formats were being developed and station owners were finally able to concentrate on content and the business aspects of broadcasting without having to worry about the chaotic problems that faced them in the early days of radio. The industry was beginning to recognize the advertising potential of radio. An hour of network program time cost around $750 in 1929. In response to the onslaught of commercial interest, the 1929 Code of the United Association of Broadcasters was introduced and provided that, after 6:00 p.m., commercial programming only of the "goodwill type" could be broadcast, and between 7:00 p.m.-11:00 p.m., no commercial announcements could be aired. Bullard also held the position of Superintendent of Radio, was the first Chief of Naval Communications, and was a delegate to the International Safety at Sea Conference in London in 1913.

Before his demise, Bullard was lobbying the House Appropriations Committee for funds to support provisions of the Radio Act. Bullard wanted to allow press associations and newspapers to use low wavelengths for point to point service and improved service.

Bullard died of a heart attack, on November 24, 1927, at 61.

By Gerald V. Flannery, Ph.D. and David A. Male, M.S.

CALDWELL, ORESTES H.
1927-1929

Orestes Hampton Caldwell was born in Lexington, Kentucky in 1888, and graduated from Purdue University. He was married to Mildred Bedard, and they were the parents of two girls. Caldwell was named by President Calvin Coolidge in 1927 to the Federal Radio Commission, the predecessor of the Federal Communications Commission. Commissioner Caldwell spent a stormy two years helping to assign frequencies during the early days of commercial radio. Caldwell fought with politicians and broadcasters during a period when the lucrative radio broadcast band was being sliced up among competing stations. Caldwell was the only engineer on the original Commission and he was assigned special supervision of the Northeast area.

The Commission found near chaos on the airways in the mid twenties, with signals from one station interfering with those from another, with little or no technical standards for equipment, and few, if any limitations on license use. To deal with that, the FRC, on November 11, 1928 ordered reallocation and divided the broadcast spectrum in a pattern that has remained substantially unchanged to this day . That effort stirred up controversy, and Orestes Caldwell, being a forceful speaker and a man with a taste for combat, was the center of that controversy.

Caldwell's confirmation as a member of the FRC was delayed for over a year because some senators complained that he was too favorably inclined toward the Radio Corporation of America, the company formed to protect America's in radio as it grew internationally. The final Senate vote giving him the job came in March,1928: it was a close 36 to 35.

During his time on the FRC, William S. Paley was elected president of Columbia Broadcast System, heading up the second network, the first being NBC with its separate red and blue networks. Later that year Dr. V.K. Zworkin demonstrated his kinescope (cathode-ray) television receiver to a group of radio engineers. Broadcasting magazine, in its 1971 chronology of radio events, listed some of the items the FRC dealt with during that time. Some of them were: the stock market crash and its effect on business; the introduction of station break announcements; the struggle between ASCAP and station owners represented by the NAB, and the fact that phonograph companies were labelling their records "Not For Use On Radio."

Commissioner Caldwell did not help soothe any of that Washington opposition when he later told engineering students at Purdue, the very next month that, "radio had become the football of politicians". He got a good reception at Purdue, though: it was his alma Mater. Caldwell told an interviewer, not long after that speech, he was thinking of resigning. He was quoted as saying: "When I first came down here, I thought we surely would have the thing cleared up in a year. We made such a mess of it during the first year that my conscience would not let me quit"

Caldwell, as one of the original members of the Commission had to face a broadcast industry that welcomed government control of certain aspects of the business but were extremely wary about other areas. Station owners were pleased to have an assigned frequency and location for their station, were pleased that the government had required certain technical standards for equipment . In essence, they were given a monopoly, a license to do business at one location in the nation and at one spot on the dial. However, they wondered where government control would end.

When Caldwell did leave the Commission in February 1929, it was to return to McGraw-Hill business publications, a place he had first worked for in 1910. At McGraw-Hill he had been co-founder and editor of publications like Electronics, Radio Retailing and Electrical Merchandising. He stayed there, after his return from Washington, until l935. Then he and an associate, Maurice Clements, formed their own publishing company, and founded the publications Electronic Industries, Electronic Technician and Mart. From 1935 to 1953 he was an officer in his own company Caldwell Clement, Inc.

Even after his retirement Caldwell remained an enthusiastic public supporter of the broadcast industry. In 1931 he staged a demonstration in a National Broadcasting Company studio. He dropped a pin from a height of six inches onto a sheet of tin, in front of a microphone, producing a sound like an explosion. Caldwell said this demonstrated "that radio had converted the whole North American continent into one vast auditorium." He said it also proved that the acoustics were so perfect that a pin dropped at one edge of the continent could be heard the length and breadth of the land.   During the 1930's, newspaper interviewers occasionally stopped by his fifty acre Greenwich farm to marvel at the electric gadgetry he had set up there. He had timing devices which turned on the lights, woke the family, and clicked on electrical devices that cooked breakfast.

Caldwell was President of the Amateur Astronomers Association, and the New York Electrical Society, trustee of the New York Museum of Science and Industry, a fellow of the American Institute of Electrical Engineers and the Institute of Electrical Engineers; he also served on the National Color Television Standards Committee and the committee on Educational Television Awards.

Caldwell died at 79 at his home in Greenwich, Connecticut on August 27, 1967 .

By Gerald V. Flannery, Ph.D. and Bobbie DeCuir, M.S.

DILLON, COL. JOHN F.
1927-1927

Lt. Colonel John Francis Dillon, one of five original members of the Federal Radio Commission, came to public service after a distinguished military career that saw him in battle in North America against the Indians and later in Europe against the Germans. While in the U.S. Army, Dillon gained expertise in the field of communications as an officer in the Signal Corps. Though his death, which would come only a few months after he was named to the FRC in 1927, cut short his contribution to the regulatory agency, Dillon still was able to provide the panel with the insight of an expert in radio communication. It was his knowledge of the field that led President Calvin Coolidge to nominate Dillon to the Commission.

Dillon was born in Belleville, Ohio, on March 6, 1866. In 1883, at the age of 17, he enlisted in the service. During his military career, he was engaged in several expeditions against the Indians. In the Spanish-American War in 1898, he served in Cuba and Puerto Rico, later going to the Philippines to help put down an insurrection there. During World War I, Dillon, now in his early 40's, was to serve as a signal officer with Company C of the 102nd Infantry during the Meuse-Argonne offensive—action for which he was later awarded the Distinguished Service Cross by General John "Black Jack" Pershing, commander of the Allied European Forces. Dillon received the honor on Oct. 14, 1918. Here's how his deeds were recorded in The New York Times of Oct. 15, 1918:

For extraordinary heroism near Chateau-Thierry, France, July 22, 1918. After being wounded, he [Dillon] refused to go to the rear, but volunteered to act as a runner and repeatedly carried messages through enemy barrages. Later the same day, he voluntarily joined a platoon and fought with it in a successful attack against the enemy's line.

After military service, Dillon became one of the country's first radio inspectors, beginning his Department of Commerce job in New York in 1919, and later working in Washington, D.C., Cleveland and Chicago; and finally becoming supervisor of radio for the 6th District, which was headquartered in San Francisco and comprised the states of California, Utah, Arizona, Nevada, and the Hawaiian Islands.

Dillon's influence in the burgeoning field of radio was evidenced in the early 1920's. The period of 1921-24 was the beginning of the broadcast era, and Dillon was to play a part. The Department of Commerce throughout this early period and on into 1927 regulated radio broadcasting under the provisions of the Radio Act of 1912. This regulation was under the direction of Secretary of Commerce Herbert C. Hoover and the staff of the Bureaus of Navigation and Standards. On Sept. 15, 1921, the Department of Commerce began to license stations as "limited commercial stations."

By the end of 1922, there were more than 570 broadcast stations licensed, and problems were developing as interference increased proportionately with the density of stations in a particular geographic area. But there were other problems also, with the technical quality of the equipment being designed and sold, with the fact that one piece of equipment wouldn't work with a competitor's brand, that the signals wandered off frequency, that some radios were sold with wet cell batteries that leaked and lost power as they aged, that transmitters were moved from their original location to some other spot deemed better to broadcast from, and more.

To solve the problem, Hoover called for the First National Radio Conference which took place on February 27, 1922. As officials contemplated various options over the following weeks of the conference, Dillon, by then a radio supervisor in San Francisco, handed Hoover a letter dated March 27, 1922, suggesting that the country be divided into zones, with different wavelengths allocated to adjacent areas; and "whenever necessary the same wavelength to be assigned to alternate zones, or to zones remotely situated." But the problems of radio were not settled at he first conference, nor the second, nor the third. It took a fourth conference before the groundwork for the Federal Radio Commission was laid.

Dillon also suggested that each station be classified, "indicated by a letter of the alphabet, according to the character of the matter which it is engaged in broadcasting." As the level of interference increased in 1922 between stations, Dillon's suggestions eventually were carried through. It would be five years later, however, while still working in San Francisco, that Dillon was finally chosen for the fledgling FRC. It was reported on March 2 of that year that President Calvin Coolidge had made up his mind on his five nominations for the new commission. Dillon, a

Republican and considered an technical expert on radio, was nominated by the president for a two-year term. Three other nominees—Henry A. Bellows, Admiral. W.H.C. Bullard, and Orestes H. Caldwell—also were considered radio experts. The proposed chairman, Eugene O. Sykes, however, was not. As The New York Times of March 2nd put it:

> The President ignored the importunities of politicians in making the appointments. He felt that the Commission should be composed of me having knowledge of radio and the appointment of four were made on the basis of their expertise in radio and electric communication.

There were some in Congress who said they would fight the confirmation of Dillon, Bellows, and Caldwell because it was feared they were too much under the control of Hoover. But on March 3, 1927, the three men were confirmed by the U.S. Senate.. The new Commission took up its duties on March 15, 1927. That was the year that CBS went on the air with a basic network of 16 stations. Dillon was the Commissioner from the FRC's Fifth Zone, however, his time would be short. By the summer of that year, Dillon was said to be ill. On June 22, 1927, The New York Times reported he had arrived in San Francisco. "suffering from an infection of the jaw," and "the crisis was expected in a few days." He died on Oct. 9, 1927, at Letterman Hospital in San Francisco. He was 61, survived by a widow and three daughters.

One week later, on Oct. 16, 1927,the FRC adopted the following resolution:

> Whereas the Federal Radio Commission has with the death of Lt. Colonel John Francis Dillon lost one of its five original members, a gallant soldier, a skilled engineer, long and intimately concerned with the problems of radio communication, a public servant devoted to the interest of the people of the United States, a tireless worker and a courteous, kindly and generous comrade, be it therefore: Resolved, that the Federal Radio Commission place in its permanent records the words: In memory of Lt. Colonel John Francis Dillon, to whom the art of radio communication in America owes an endless debt for his wise counsel, his clear vision, and his devoted labor as the first member of the Federal Radio Commission from the Fifth Zone.

By Michael A. Konczal, M.S. and Gerald V. Flannery, Ph.D.

HANLEY, JAMES H.
1933-1934

James Hugh Hanley served one of the shortest tenures on the short-lived Federal Radio Commission -- less than two years.

Hanley's nomination was seen as a political favor from President Franklin D. Roosevelt given for Hanley's support for the president and the Democratic Party in the mid-west; in fact, he was one of the first people to travel around the country organizing Roosevelt for President clubs in 1929 and 1930. His nomination was supported by Arthur F. Miller, The Democratic National Committeeman for Nebraska, his law associate and Roosevelt's first choice for the Commission nomination. The nomination was also supported by Postmaster General James Farley. Hanley was chosen and confirmed over President Herbert Hoover's nominee, J.C. Jensen.

Hanley's lack of experience in broadcasting, and the fact that he was only a "casual listener" of radio, contributed to the nickname "Rookie" he was given. In his confirmation hearings before the Interstate Commerce Committee, Hanley said that radio meant no more to him than it did to the average man on the street. He said that he had no experience in radio, but he had an open mind toward the industry.

James Hugh Hanley was born in O'Neil, Nebraska, on July 4, 1881. He graduated from Freemont Normal College in 1903 and served as a principal in the Nebraska school system from 1903 -1907. In college he played football and baseball.

After completing a law course at Creighton College of Law in 1910, he began practicing. From 1911 to 1919 he served as secretary to Representative C.O. Loebeck (D-Neb.). From 1918 to 1922, Hanley practiced in Nebraska, and from 1922 to 1924, he was appointed as the first Nebraska Prohibition Director by President Woodrow Wilson. Also in 1922, Hanley was the Democratic candidate for Congress from the Second Congressional District. In April, 1933, he went to Washington, D.C., to serve on the Federal Radio Commission. He stayed there until 1934 when the Commission was dismantled. From 1935 to 1941,

Hanley practiced radio law in Washington. In 1941, he was named to the Board to Adjucate Cases for Conscientious Objectors. Hanley served as a special attorney with the anti-trust division of the Justice Department until shortly before a heart ailment caused his death on July 9, 1945.

Hanley believed that the American system of broadcasting was the best in the world, but that certain practices and abuses, that discredited and possibly disrupted the whole institution, had crept into the system. He was interested in educational programming and educational stations and thought they should be handled liberally and that they should be allocated more channels.

Hanley's viewpoint on the role the Commission should play in regulating radio was brought out in his confirmation hearing before the Interstate Commerce Committee. Hanley stated that since he was a Democrat, he was opposed to monopoly of all kinds. He said that radio was a very strong industry that could lead to a monopoly. Hanley also said that he recognized the right of the government to regulate the radio industry and safeguard the rights of the people.

After six months on the Commission, Hanley told Broadcasting magazine that he was convinced that the listener's viewpoint should be more of a factor than it was then in program-building. In fact, he suggested that listeners write to the Commission about their likes and dislikes and that they would be weighed in decisions. Furthermore, Hanley stated that he would like to have questionnaires sent to leading citizens, public officials, educators, and other representative groups to get a cross-section of listener opinion.

Hanley thought that banning provocative broadcasters such as those who incited religious unrest and other strife from the airwaves was the duty of the Commission. The Brinkley case (KFKB Broadcasting Association, Inc., v Federal Radio Commission) involved a medical malpractitioner who diagnosed ailments over the radio and offered his own prescriptions. The case was the first judicial affirmation of the Commission's right to consider a station's past programming with relation to "public interest, convenience, and necessity" when license renewal was sought. Shuler (Trinity Methodist Church, South, v Federal Radio Commission) involved Reverend Shuler and his defamatory and objectional utterances over the station. The D.C. Circuit of Appeals built on Brinkley and ruled that refusal to grant license renewal violated neither the First nor the Fifth Amendments.

Hanley also supported censorship of material that was not deemed appropriate for representative groups of people. A number of other things came before the Commission during his two years on board. The press-radio war was underway and the Associated Press membership voted to ban network use of its news on radio and further told radio stations they would have to pay to have their program logs published. The repeal of prohibition raised the question of whether there would be liquor ads on radio, and Mexico scuttled the First North American Radio conference by demanding that twelve clear channels be set aside for it.

After a year on the Commission, Hanley remarked in The New York Times that there was too much concentration of facilities in the hands of a few who had found it financially advantageous to use them in populated areas. He pointed out that in rural and farming areas there was a dearth of such facilities.

Hanley thought that there were too many clear channels and that more people would have been served and service would have been improved, especially in sparsely populated areas, if there were more regional channels. He suggested a compromise or merger of broadcasting stations to eliminate duplication of programming and to make the limited number of existing channels be distributed most beneficially to broadcasters and the public.

Hanley's term as Commissioner ended early, July 10, 1934, when the Communications Act of 1934 rendered the FRC obsolete and replaced it with the Federal Communications Commission. During the last 14 days of the FRC's reign, it granted 150 applications for increases in day and night powers, changes in frequencies and, in fact, every request that could be granted without a hearing. In many cases, applications were approved within twenty-four hours.

Some of the things the FRC had to deal with during his term were placing the full facilities of radio under Roosevelt's temporary control during the banking crisis of 1933, the fight between the American Society of Composers and Publishers (ASCAP) and the National Association of Broadcasters (NAB) over royalties and the use of music on the air. That same year the U.S.Supreme court upheld the power of the FRC to exclusively grant radio licenses, and the problem of liquor advertising on radio arose after the repeal of prohibition. The radio-press war began in this time period, and the Federal Trade commission decided to review all commercial radio copy in a survey of advertising.

The new Commission was recommended by President Roosevelt in an effort to gain regulatory power over transportation, power, and communications. The FRC regulated radio; Roosevelt wanted to regulate all interstate and foreign commercial services which relied on wires, cables, or radio as a medium of transmission.

President Roosevelt was offering new recovery programs to try to bring America out of the Great Depression and at the same time centralize control of some of the diverse federal agencies and responsibilities. The Federal Government had authority over transportation in the Interstate Commerce Commission and authority over power in the Federal Power Commission. The Federal Communications Commission was proposed to give the it the authority and power to regulate the communications industry. The FRC was limited in its authority and the nation was facing the rapid growth of radio, plus the early experimental days of television, plus the refinement and further development of the telephone and telegraph.

Hanley was there when it was all being created.

By Gerald V. Flannery, Ph.D. and Brian P. Atkinson, M.S.

LAFOUNT, HAROLD A.
1927-1934

Harold Lafount came to the United States in 1893 from his native England. He attended Utah State Agricultural College where he received a degree in civil engineering. Lafount became manager of the Pacific Land and Water Company in Salt Lake City after assisting his father in the retail hardware business in Logan, Utah. He was receiver at the Sevier River Land and Water Company from 1923 until President Calvin Coolidge appointed him a Federal Radio Commissioner in 1927 thus he was one of the original Commissioners charged with regulating radio. Lafount, representing the Far Western zone, which includes the Pacific states and Hawaii, urged reducing the number of radio stations. He was against having too many small stations that could lower the chance of larger stations being heard effectively. He definitely wanted to raise the power of the larger stations. Lafount was a pioneer in improving radio reception for listeners and urged the establishment of citizens' advisory boards to help plan programs for unsponsored time on stations in small communities. During his period on the Commission, ASCAP, the American Society of Composers and Publishers battled radio station owners over rights and royalties due their members for music and lyrics played over the air. It was also a time when the stations were turning their efforts to improving program fare introducing many of the formats that later became standard on radio and then television. He was also one who pushed for the perfection of television and in 1931 he proposed censoring television. Lafount believed that through censoring television programs that abused the medium through distasteful advertisements or through objectionable programming, the Commission would be able to enhance television and its uses for the future.

Lafount was there when the first issue of <u>Broadcasting</u> magazine was published on October 15, 1931. It contained a report on a speech by Walter J. Damm, president of the National Association of Broadcasters, who said:

Broadcasting in the United States today stands in grave jeopardy.

Politically powerful and efficiently organized groups, actuated by selfishness and with a mania for power, are now busily at work plotting the complete destruction of the industry....

Lafount took part in the six month investigation of advertising on radio that concluded "any plan...to eliminate...commercial advertising...will, if adopted, destroy the present system of broadcasting." He was also instrumental in drawing up a set of rules for political radio programs. In August 1932, the Republicans allocated $300,000 to buy time on radio for the presidential election. Although he was a Republican, he refuted a 1933 Republican charge that the Roosevelt Administration sought to censor radio broadcasts. He left the FRC in 1934 to take charge of all Bulova radio interests.

Later, Lafount became the president of the Atlantic Coast Network in New York, the Broadcasting Service Organization in Boston and the National Independent Broadcasters. He vice president of the Wodaam Corporation, the Greater New York Broadcasting Corporation and the Fifth-Forty-Sixth Corporation.

Lafount was married and had four children, all daughters. He died October 21, 1952, at the age of 72 at the home of one of his daughters, Mrs. George Romney.

By Gerald V. Flannery, Ph.D. and Ricky L. Jobe, M.S.

PICKARD, SAM
1927-1934

In January 1927, the National Association of Broadcasters called for the formation of a Bureau of Radio Communication In the Department of Commerce on the grounds of economy and efficiency. Its members would serve on the Bureau only after passing a Civil Service test. This would ensure that Bureau memberships were independent of the influences and considerations of politics and politicians. In 1927 Congress created the Federal Radio Commission with the Radio Act of 1927. On March 2,1927, President Calvin Coolidge appointed the original five members of the FRC, one of them was Sam Pickard.

Pickard was only thirty-one years old when he accepted his appointment on November 1,1927 after the resignation of Henry A. Bellows.That made him one of the youngest Commissioners to ever serve on the FRC. Pickard had been Secretary of the FRC since its inception. That experience made him a good candidate to fill the vacancy. He also had been an educational broadcaster In Kansas and for the Department of Agriculture. He started a "college of the air" at Kansas State Agricultural College and broadcast farm programs to over 100 stations in the midwest for the U.S. Department of Agriculture. He was described In a statement by the Commission on the day of his appointment and printed in The New York Times as "one of the outstanding figures In the radio world because of his pioneer work, especially among the farmers of the Middle Western United States."

He was once Extension Editor at Kansas Agricultural College. Later, Pickard moved to Washington to head the radio service for the Department of Agriculture. Pickard was responsible for broadcasting farm programs. His "farm school of the air" made his name known in Washington as an innovative and often inspiring force In the infancy of educational radio. The programs became such a success that many farmers considered their radio as their most important piece of farm equipment.

During his time on the FRC, Pickard was In charge of the Fourth Zone In the midwest. There was a rule stating that each Commissioner must reside In the zone which they were allocated to control; being from Kansas, Pickard was the likely choice.

One of the most interesting statements made by Pickard during his tenure as the official spokesperson for the FRC was the announcement that an effort would soon be made to stabilize the broadcasting situation by the issuing licenses good for a longer period of time. The limit was sixty days in the beginning months of the FRC. He also announced that 300 stations would be eliminated by the commission on February 1, 1928, when all licenses expired. There were some 690 stations In existence then. Commission policy at this time was "survival of the fittest" to clear airwave interference. Pickard stated that survival and license renewal would depend on priority of existence and public service rendered. A year after these changes were carried out, Pickard reported clearer reception in the Midwest. He said that there were more stations on the dials that had stronger reception. He was reappointed on March 30, 1928, at a time when the FRC's main goal was to put the Radio Act of 1927 Into successful operation.

Also at this time In history the first visual images were being transmitted over the airwaves. Pickard made a statement to The New York Times that it was not feasible to not buy a radio and urged people to just walt until television had become a reality. He said it would be several years until television was practical. Television, in 1928, was only on the threshold of picture transmission. In another statement, made to The New York Times, Commissioner Pickard predicted that the phonograph record would be the answer to broadcasting's financial problems. He also felt that government supervision should be minimal In broadcasting.

During his last year with the FRC, Pickard s most newsworthy accomplishment was his assistance with the national broadcast of President Herbert Hoover's inauguration. He flew in the navy plane that sent the signal and for the first time In history, the proceedings of the Senate (Inauguration) were broadcast over the airwaves. In his younger days, Pickard was a pilot in World War I. He was shot down in France on October 31, 1918, where he almost lost his life. Hoover's inauguration was the first time he had flown since that near tragedy.

Broadcasting magazine, in its 1971 chronology of important events, listed some of the items the FRC dealt with during his tenure. They were: the presence and growing influence of both the NBC and CBS radio networks; the use of advertising on radio; the stock market crash and its effect on business; the election of William S. Paley as president of CBS; the demonstration by Dr. V.K. Zworkin of his kinescope or cathode-ray television receiver; the first issue of Broadcasting magazine; the introduction of station break announcements; the struggle between ASCAP and station owners represented by the NAB; and that phonograph companies were labelling their records "Not For Use On Radio." Two out of every five American households had a radio, despite the depression; radio placed all its facilities at the disposal of President Franklin D.Roosevelt during the banking crisis; and Associated Press members voted to ban network broadcasts of AP news.

After a private meeting with Hoover, Pickard resigned his position with the FRC on January 30, 1929. His resignation came after the Columbia Broadcasting System made him an offer of Vice President. Hoover then nominated Major General Charles McKinley Saltzman to fill the FRC vacancy created in Pickard' s 4th zone.

After his years as Vice President of CBS, Pickard was again a public figure when his name appeared in the papers because the Federal Communications Commission refused to renew the license for radio station WOKO in Albany, New York. The renewal was refused because Pickard concealed the 240 shares of stock he had owned in the company for 12 years. On December 10, 1946, the U.S. Supreme Court upheld that refusal by the FCC. Eight Justices concurred in the decision, with no dissent. The court ruled that the Commission, not the court, must be satisfied that the public interest would be served by license renewal. The Supreme court would not substitute its judicial discretion for the administrative authority of the FCC. Sam Pickard reportedly was sold the 240 shares after his assurance that he would help to secure a CBS affiliation for station WOKO; also that he would furnish, without charge, Columbia engineers to construct the station, would supply it a grand piano and certain news publicity about the opening of the station. Many innocent stockholders suffered consequences as result of Pickard 's stock concealment.

By Gerald V. Flannery, Ph.D. and Carla P. Coffman, M.S.

ROBINSON, IRA E.
1927-1932

Ira Robinson never liked the idea of recorded music on radio. He felt the general public could listen to phonograph records anytime at home but they wanted to hear live music on their radios. "I'm wondering," he said, " whether or not all this great spectrum covering the United States of America . . . is not too great a thing to be devoted to the reproduction of phonograph records."

Robinson, whose only radio expertise was a talent for picking up long-range stations on his radio at home, was appointed by President Calvin Coolidge on March 29, 1927 to represent the Second Zone on the newly-established Federal Radio Commission. Robinson replaced the late Rear Admiral W. H. G. Bullard. The Commission selected Robinson as its Chairman shortly after his Senate confirmation. It was rumored that Coolidge asked between twenty and thirty people to accept the position before Robinson agreed.

Born September 16, 1869 in Grafton West Virginia, Robinson studied law at the University of West Virginia. He was admitted to the Bar in 1891. The following year, he married the former Ada Sinsel, who died in 1930. She bore him two children, a girl and a boy. Robinson's son died in infancy.

He was a member of the West Virginia Senate for two years before he was appointed to that state's Supreme Court in 1907. He served as President, or Chief Justice, of the Court from 1909 until he stepped down from the bench to run for governor as the Republican candidate. He also was past president of the Institute of Criminal Law and Chairman of the criminal law section of the American Bar Association.

Before being chosen to serve on the FRC, Robinson was in great demand by many prominent law schools as a lecturer. While still a Justice, he taught at the University of West Virginia School of Law from 1913 to 1920. He was also the 1919 Hubbard lecturer at the Albany School of Law and was invited to speak at Northwestern University in 1920.

After World War I, he adjucated War Minerals Claims for the

Department of the Interior. Before his nomination to the Commission, he was negotiating a deal for the government to purchase the Cape Cod Canal.

One of the first tasks put before the Commission was the reallocation of power to radio stations as mandated by the Davis Amendment. At the time, there were so many stations in existence that they had to share the limited cleared channels available. This caused the airwaves to be badly cluttered. The FRC proposed to solve this problem by reducing the number of stations in operation and increasing the stations' power output. It was a controversial proposal.

Robinson refused to sign the reallocation plan. He felt that rather than take stations off the air, new stations should be granted licenses in the areas that were being underserved. He also thought that his region, the Second Zone, was not adequately covered. The plan went into effect anyway without his signature.

In an article in the January 20, 1929 edition of The New York Times, Robinson wrote that since the reallocation plan was now in effect, the FRC should turn its attention to other industry problems. He felt the board should adopt rules and regulations governing radio technology and the type of equipment used, station locations, broadcasting chains, and the use of stations by political candidates. He thought regulation was needed to protect listeners from fraudulent advertisers and the unfair advertising policies of some station owners. He also wanted the FRC to explore the new radio technology being developed in order to decide how to regulate it.

In March, 1929, Robinson entertained the idea of resigning from the FRC after receiving several offers from the private sector. According to The New York Times, Robinson was disappointed that others on the board did not share his opinion that short wave licenses should only be granted to those who would use them for the good of the public. His belief that the airwaves were a public utility guided all his decisions while on the FRC. In an address to a Washington, D. C. conference on the use of radio in education, he said, "Radio was born a crippled child, birthmarked by advertising, and it is up to us to see that it gets out of that."

In April, 1929, Robinson argued that both sides of a debate should receive equal radio time. The Chicago Federation of Labor, operating station WCFL, wanted to increase its power and to be given permission to broadcast at night in order to better reach laborers who worked during the day and were unable to hear the station's programs. Robinson reasoned that if the union could use WCFL to broadcast propaganda, the business owners should have the same rights as well.

He suggested that the government should regulate stations in order to assure that "educational and high-class programs" get a guaranteed amount of airtime per day. Robinson's idea of "high-class programs" included those that were aimed at farmers and other country folk. These programs, he felt, should not be cast aside in lieu of more sophisticated "city" programming.

Robinson was also the first to suggest that the government charge a fee for broadcast licenses. The money collected could be used to offset the administrative costs of the Commission.

Those seeing Robinson's affinity for long-range stations as a sign that he would be in favor of more cleared channels were soon disappointed when he voiced his opposition to such a request. A group of ten stations asked for an increase from forty to fifty cleared channels and the adoption of better regulations to govern the use of those channels by regional and local stations. At that time, stations sharing channels had to alternate the days and nights that they could broadcast. This left several cities without radio service fifty percent of the week. The group argued that the new channels would alleviate this problem and would benefit radio fans in rural areas as well. Robinson was unconvinced.

He turned over Chairmanship of the Commission to Charles M. Saltzman on March 1, 1930, and then became a liaison between the FRC and other government agencies with an interest in radio. Robinson retired from the Commission on January 15, 1932 and returned to his private law practice. In 1936, he went into partnership with his nephew Rupert A. Sinsel. Nine years later, they formed the law firm of Robinson, Sinsel, and Swiger in Clarksburg, West Virginia.

Robinson had always been involved in the Methodist-Episcopal church. He served as a delegate to the church's general conferences in 1912 and 1916. He was later a member of the committee to unify the three branches of Methodism from 1939 to 1940.

In April, 1951, at the age of eighty-two, Robinson married for the second time to Loretta E. Malone. On October 28, just six months after their wedding, Robinson died in Phillippi, West Virginia.

By Robyn Wimberly Blackwell, M.S. and Gerald V. Flannery, Ph.D.

SALTZMAN, CHARLES M.
1928-1932

Charles McKinley Saltzman was born in Panora, Iowa, on October 18, 1871. After high school he entered West Point Military Academy, where he graduated in 1896. Saltzman then entered the cavalry. After the war with Spain he returned and married Mary Peyton Eskridge of Boston. He was then appointed signal officer on the staff of General Leonard Wood in the Philippines, where he served from 1901 to 1905. This was followed by three years as an instructor in the Army Signal School at Fort Leavenworth, Kansas, then five years in the electrical division of the Signal Corps in the War Department in Washington. Saltzman was ordered to Panama as signal officer in 1915 and installed the radio, telephone and telegraph systems there. Just before the United States entered the first World War, Saltzman was recalled to Washington as executive officer in the office of the Chief of the Signal Corps, of which he was himself made Chief in 1924, holding the post until his retirement. For his war-time services he received the Distinguished Service Medal in 1919.

A year after his retirement from the Army in 1928, after thirty-two years of service, Major General Saltzman was appointed to the Federal Radio Commission. He became its Chairman from 1930 to 1932 when he resigned. The radio industry applauded his appointment to the Commission, for he had "grown up" with radio, and was the only member of the Commission drawn from the ranks of expert radio men. General Saltzman was one of the best friends of the American radio amateur. He conceived and organized the "army amateur" network which, in 1926, included 2,000 amateur radio stations throughout the United States. He formed it as an emergency chain with an understanding that each young operator, in the event of war or other emergency, would place the station and operator at the disposal of the government.

The association of General Saltzman with modern methods of communications began in his boyhood when he was asked to take over the key of an absent wire operator at a railroad station. That began a career where he worked as a telegrapher for railroads and the Western

Union before going to West Point. He was a delegate to the Radio Conference in London in 1912, the International Telegraphic Conference at Paris in 1925, the Radio Conference in Washington in 1927, and the Radio Technical Conference at The Hague in 1929. The broadcast industry flourished during the Depression era. It was one of the few industries that did. In October of 1931, as Chairman of the Federal Radio Commission, General Charles McKinley Saltzman called the American broadcasting industry "the best in the world."

Broadcasting magazine, in its 1971 chronology of radio events, listed some of the items the FRC dealt with during his tenure. They were: the presence and growing influence of both the NBC and CBS radio networks; the use of advertising on radio; the stock market crash and its effect on business; the demonstration by Dr. V.K. Zworkin of his kinescope or cathode-ray television receiver; the first issue of Broadcasting Magazine which covered radio and the FRC in depth; the introduction of station break announcements; the growing tendency of owners to develop a station and then sell it; the struggle between ASCAP and station owners represented by the NAB; and the fact that two out of every five American households had a radio.

McKinley Saltzman died in Walter Reed Hospital on November 25, 1942 at 71.

By Gerald V. Flannery, Ph.D. and James Nunez, M.S. candidate.

GARY, HAMPSON
1934-1935

Hampson Gary served for a short period of time as a Federal Communications Commissioner -- in fact, he was an original member of the panel, which succeeded the old Federal Radio Commission in 1934.

Gary was born in Tyler, Texas, on April 23,1884. Educated at Bingham School in North Carolina, he earned his law degree from the University of Virginia. Admitted to the bar in 1894, Gary practiced law in Tyler. In 1921, he moved to Washington, D.C. and later practiced in New York also.

In 1901, Gary married Bessie Royall of Palestine, Texas, and the couple had two children, Franklin and Helen. An Army veteran, Gary served as a captain in the Spanish-American War with the United States Volunteers. He was later a colonel in the 3rd Texas Infantry. Following the war, Gary won a seat in the Texas House of Representatives, 1901-02. He also served as a member of the Board of Regents for the University of Texas for several years. In 1914, Gary was named referee in Texas bankruptcy court for four years and standing master in chancery for the U.S. District Court for two years. Also in 1914 he was appointed special counsel to the U.S. Department of State to help in matters arising out of World War I. He became a solicitor for the department in 1915, and in 1917, he became consul general to Egypt, a position he held until 1920. In 1919, he was called to France to work with the American Commission to negotiate the Peace of Versailles ending World War I. Gary then went to Switzerland in 1920 as a diplomat, staying there for one year before returning to the United States. He practiced law in Washington, D.C. until 1934, when he was nominated by President Franklin D. Roosevelt to the new FCC, which began its work on July 11, 1934. Gary became Chairman of the FCC's Broadcast Division. As to his philosophy about the role of the FCC, Gary was once quoted as saying, "We have one of the finest systems in the world. We do not want to exercise bureaucratic control and we do not want to dictate... entertainment or discussion ... on the air."

Gary was appointed to the Commission for one term, but privately had agreed to stay only until January 1, 1935. The New York Times, on

January 3, 1935, reported it :

> The resignation of Hampson Gary as a member of the FCC was
> announced a the White House today, and ... Anning S. Prall,
> former representative from New York, has been appointed to
> fill the vacancy... He has been mentioned as a probable
> appointee to a diplomatic post.

Gary, however, did not go back to diplomatic life, but instead became
general counsel for the FCC. During the next three years, he tried to get
an appointment back on the FCC, but was not successful. In late 1938,
FCC Chairman Frank R. McNinch, demanded Gary's resignation. Gary
refused. The scrape was reported in the October 13, 1938 edition this
way: "Officials of the FCC were reluctant to discuss the reasons for the
action of McNinch, who returned to duty today after an illness of several
months. McNinch did not give any reason for asking for Gary's
resignation."

The situation became somewhat clearer after Oct. 13, 1938, when on
that day the Commission held a meeting and voted 4-2 to dismiss Gary
as its general counsel.

Again The New York Times reported it:

> In a move toward a "clean-up" ordered more than a year ago by
> President Franklin D. Roosevelt, the FCC today ousted Hampson
> Gary of Texas as its general counsel, supplanting him by William J.
> Dempsey, youthful protege of Thomas G. Corcoran. Mr. Gary's
> dismissal was ordered by a 4-to-2 vote of the Commission, upon a
> resolution by Chairman McNinch... that it was "necessary for the
> proper and efficient discharge of the function of the commission to
> change its general counsel.

The only accusation against the dismissed counsel was made
informally by McNinch, who said that Gary's "inefficient management " of
the law department, and his "lack of administrative ability properly to
direct the work of the department" had prompted the resolution."

The major significance of the action arose from the clear intimation by
Chairman McNinch that it was the first in a series designed to rearrange
the personnel of the Communications Commission, and from the fuel
which dismissals will add to the demand for a Congressional
investigation of this problem child of the Administration.

It was rumored at the time that Gary might take up the job of general
counsel of the Reconstruction Finance Corporation, but although the

RFC's Chairman, Jesse Jones, said he would hire Gary, the attorney apparently never took the job. Asked about the controversy at the time, Jones told <u>The New York Times</u> on October 19, 1938, he felt the removal of Gary was "just a case of two fellows not getting together, and "sometimes good men just don't get along."

Gary returned to private life after his dismissal from the FCC and continued to practice law in the Washington area until his retirement. He died on April 18, 1952, just days shy of his 78th birthday, at Good Samaritan Hospital in Palm Beach, Fla. He was survived by his two children.

By Michael A. Konczal, M.S. and Gerald V. Flannery, Ph.D.

STEWART, IRVIN
1934-1937

Over three decades of service in numerous positions related to national communication policies, qualified Irvin Stewart as one of America's authorities on the mass media. His career in the communications field began in 1927 when he served as a member of the American Delegation to the International Radio Conference in Washington, D.C. At the age of sixty-four, he served as a Director of Telecommunications Management for President Kennedy's Executive Office from 1962 through 1963. In between these thirty-six years lies an interesting history of a man of many accomplishments.

Stewart's educational background provided him with ample credentials to serve on the 1927 conference when only twenty-eight years old. He received both his Bachelor's and Master's degrees from the University of Texas in 1922 and a Ph.D. from Columbia in 1926 in the areas of law and government. He went on to teach government at the University of Texas from 1922 to 1929 while simultaneously serving as an Assistant Solicitor to the United States Department of State from 1926 to 1928. With such experience behind him, Stewart became head of the Department of Government at the American University Graduate School in Washington from 1929 through 1930. Now in the center of American politics and with his energetic personality, Stewart was put in charge of electronic communications for the Treaty Division of the Department of State from 1930 to 1934. In this capacity he served as a member of the International Technical Consulting Committee on Radio Communication, Copenhagen, 1931; Pan American Commercial Conference, Washington, 1931; International Radio Conference, Madrid, 1932; International Telegraph Conference, Madrid, 1932; North and Central American Regional Radio Conference, Mexico City, 1933. During this period he was also a member of the Interdepartmental Committee on Communications working with Congressional committees to draft the Radio Law which eventually created the Federal Communications Commission. In July of 1934 he was appointed a FCC Commissioner.

During his three year term on the FCC, Stewart served as the

Chairman of the Telegraph Division (1934 - 1937) and as Vice Chairman (1935 - 1937). Through his persistence, Stewart was able to initiate a merger between Western Union and Postal Telegraph Companies. In his opinion such a move would protect the telegraph industry from the outside competition of air mail.

Stewart was also an outspoken proponent for the audience. In a speech at Duke University in 1937, he stressed the necessity for fair allocation of wave lengths and greater variety in radio programming.These were points Stewart covered in his initial work on the Communications Bill and as such the FCC was given the authority to determine radio distribution from state to state.

Stewart joined the FCC in its maiden year as the new agency took over from the FRC and tried to deal with the remaining problems in communications. The FRC had handled the assignment of radio frequencies, tower locations, call letters and time of operation. It had set standards for the technical equipment that could be used in the broadcast industry, but now the range of things covered by the FCC was far greater.

In 1937, following the expiration of his FCC term, Stewart became Director of the Committee on Scientific Aids to Learning until 1946. He also served as Executive Secretary to the National Defense Research Committee (1940 - 1945); the Committee on Medical Research (Office for Emergency Management) (1941-1945); Deputy Director of the OSRD (1946). Stewart returned to his educational pursuits as President of West Virginia University at Morgantown from 1946 to 1958. He was awarded further degrees from Waynesburg (1946), West Virginia Wesleyan (1948), West Virginia State College (1948), Marshall College (1953) and Bethany College (1954). From 1958 to 1967, he served as a Professor of Government at West Virginia University.

Government service remained prominent in Stewart's life. In 1950, President Harry Truman established a Communications Policy Board to study domestic and international problems such as the scarcity of radio frequencies and government communication facilities. Stewart became Chairman of that committee.

The 1950's continued to be busy years for Stewart with work on the National Committee for Developing Scientist and Engineers (1956), National Advisory Committee for Rural Defense (1956), Chairman for the American Delegation of International Conference on Organizational and Administrative Applied Research (1956), and the West Virginia Constitutional Revision Committee (1957 - 1962).

Stewart went on to serve President Kennedy as his Director of Telecommunications from 1962 -1963.   The position carried the responsibility of coordinating and formulating policies for telecommunications activities in the Executive Branch of government . In 1967 he served as a consultant for the National Academy of Public Administration. In 1948 he received the Presidential Medal of Merit. Dr. Irvin Stewart, born in October 1899 in Fort Worth, Texas was married to Florence Degendorf and had a son, Richard Edwin. Over the course of his eighty-odd years, he acquired quite an impressive list of accomplishments both in the public and private sectors. He served the government in numerous capacities for over three decades, taught government and law over forty years, and wrote about these topics. Stewart's mark in the world is well recorded.

By Peggy Voorhies, M.S. candidate and Gerald V. Flannery, Ph.D.

PAYNE, GEORGE H.
1934-1943

George Henry Payne brought a combination of law, journalism and political savvy to the Federal Communications Commission when he took his seat in 1934. He zealously fought to keep radio free of commercial monopolies and government control. In speeches made at Harvard, Cornell, and Columbia Universities, shortly after joining the FCC, Commissioner Payne often criticized commercial broadcasters for their attempts to "make the Commission a subservient instrument to commercial radio." To guard against commercial broadcasters becoming too influential with the FCC, Payne brought charges against the radio lobbyists to the House Rules Committee. He claimed that free competition was disappearing in the broadcast field and that two or three major chains (NBC, CBS, and MBS) controlled or owned most of the best facilities on the airwaves. The FCC had previously rejected Payne's resolution for a Congressional investigation, and in it's ruling the House Rules Committee stated that Payne had failed to substantiate his charges. A Congressman went so far as to demand Payne's resignation to restore public confidence in the FCC.

Commissioner Payne's main concern was the quality of radio programming. At a national conference on educational broadcasting, he stressed that standards of radio programming must be improved. In his opinion, the average program of the time was addressed to a child of 12 and would perpetuate mental immaturity in adults.

George Payne was born in New York City, August 13, 1876, son of Katherine and George C. Payne. He attended New York City College for three years and studied special courses at the College of Pharmacy and was later admitted to the Law College of New York University. At the age of seventeen, and still a student, Payne became the proprietor and publisher of The Long Summer Session in New Jersey. Ten years later he began a four year term as drama and music critic for the Evening Telegram. After a two year hiatus from journalism, he joined the staff of The Evening Post in 1909 as a political reporter, remaining in that position until 1912. During this period he authored perhaps his best known book-- The Birth of the New Party, a study of the origins of the progressive movement.

After leaving the Post, he became one of the managers for Theodore Roosevelt's Presidential Campaign during 1912. The following year, he was a campaign manager for a candidate for the presidency of the New York Board of Aldermen. From 1913 through 1920, Payne was involved in numerous aspects of politics, both locally in New York and nationally. He was a lecturer on history and development of American journalism; member of the New York County Republican Committee; candidate for the New York Assembly; a manager of the literary bureau for Henry Stimson's gubernatorial campaign; and delegate to the 1920 Republican National Convention in Chicago. He also was an unsuccessful candidate for the United States Congress. For approximately 15 years prior to being named to the FCC, George Payne was City Tax Commissioner for New York City. Payne's tenure on the newly formed FCC began July 11, 1934, when the FRC was eliminated by the Communications Act of 1934. His term was due to expire on June 30, 1936. During 1936, President Franklin D. Roosevelt submitted Payne's name for reappointment for an additional term beginning July 11, 1936. During his second term, he served as Vice Chairman of the Telegraph Division of the Commission. He generally sided with the Democratic majority, even though he was normally a Republican.

Payne's second term appeared to be very controversial. It was during this period that he made his appearance before the House Rules Committee and was attacked by Representative Cox for severely damaging public confidence in the FCC. He also filed suit for libel against Broadcasting magazine during this period. Broadcasting denied the charges on the grounds that all statements were true and were fair commentary. They added that the article about Payne was of great public interest and was made in good faith, without malice. In July, 1943, President Roosevelt again submitted Payne's name for a third term on the Commission. This third term was to be for seven additional years. The nomination, however, was withdrawn the day after it was submitted to the Senate for confirmation without any explanation. The withdrawing of Payne's name remains a mystery. Following the completion of his second term, Payne returned to the public sector to become Vice President and Director of the French Telecommunications Company in New York City. George Henry Payne died on March 3, 1945, at the age of 68, at the home of his daughter in Queens, just a year and a half after his term at the FCC ended.

Payne's additional accomplishments include many books including

England, Her Treatment of America (a history of diplomatic relations between England and the United States since 1789), A History of Journalism in the United States, and A History of the Child in Human Progress. He belonged to numerous clubs and organizations including the Metropolitan and Army and Navy Clubs in Washington, the Hardware Club in New York, and the Circle Interallie in Paris, France.

By Gerald V. Flannery, Ph.D. and Mary Syrett, M.S.

CASE, NORMAN S.
1934-1945

Norman Stanley Case was born on October 11, 1888, in Providence, Rhode Island. In 1904, he enrolled at Brown University. He joined Delta Upsilon fraternity while at Brown and earned his Bachelor of Arts degree there in 1908. Electing to postpone his next career decision, Case spent a year travelling around the world to further his knowledge of people, social customs, and language. When he returned in 1909, he knew what he wanted to do, and entered Harvard Law School. Case studied law at Harvard for the next two years. His last year of law school he spent at Boston University, graduating with honors in 1912. He applied for and was admitted to the Rhode Island bar in 1911, and the Massachusetts bar in 1912, although he predominantly practiced law in Rhode Island. Back when Case had returned from his year of travelling around the world, he took it upon himself to enlist as a private in the Massachusetts National Guard. Upon graduating from Brown University with his LL. B. in 1912, Case rose to Second Lieutenant, then First Lieutenant the following year.

In 1915 he transferred to the Rhode Island National Guard, where he served on the Mexican border in 1916. After his return from the Mexican campaign, Case married Emma Louise Arnold of Bethel, Vermont. In 1917, with the rank of captain, he went to France to take part in World War I. He served with the 26th Infantry Division. Soon after his return from the war, Captain Case resigned his commission and was honorably discharged in 1919. Case was a member of the Soldier Bonus Board of Rhode Island from 1920 to 1922, and in 1923 he was admitted to the Supreme Court bar where he practiced law.

In 1927, while serving as United States Attorney for the District of Rhode Island, he was elected lieutenant governor of the state of Rhode Island. Case became the state's governor a year later when the incumbent Governor Potheir passed away. Case performed his duties well and was elected governor in his own right in 1929 and held that post until 1933.

During his term as governor, Case managed to reduce Rhode Island's state debt each year--something no governor before him was able to do. He was the first governor to appoint superior and district court judges. Governor Case also passed an Unemployment Relief Act, and saw to it that Rhode Island took care of its own relief problems.

In 1930, Governor Case received an LL. D. from Manhattan College. He received an LL. D. from Rhode Island State College the following year.

In 1934, he was appointed a founding Commissioner in the newly-created Federal Communications Commission by President Franklin D. Roosevelt. Commissioner Case served in the broadcast telephone and telegraph divisions while the FCC was being organized. He dissented on many issues, but especially on matters of policy. Norman Case was also Chairman of the Agency Committee and of the Committee of Revision of Domestic Rate Structure of the Western Union Telegraph Company. Case served on the FCC a total of eleven years, retiring on June 30, 1945.

The period of the thirties was characterized by an increasing emphasis by broadcasters on creating program formats, comedies, dramas, talk shows, big band presentations, news programs and the like. The Great Depression swept the country and President Roosevelt launched many new federal programs to deal with the dormant economy. Roosevelt also took to radio and began what became known as his famous "Fireside Chats," where he tried to speak to the American public directly in a friendly, conversational manner that lent itself well to the growing medium of radio. Despite the fact that money was short, much of America found enough to buy a radio and, thanks to the low cost of electricity, turn in on every day.

The FCC saw the growth of the Press / Radio War between members of the Associated Press and station owners, at odds over taking wire service reports of news events, and putting them on radio before they were printed / distributed in the newspaper. That battle was finally settled as newspapers moved away from trying to present the "scoop" or the "first news," and decided to provide "in depth" reports and sidebar stories, something radio was not equipped to handle. Then World War II came along and radio went to war; reporting changed, commentary crept in, and Edward R. Murrow and Lowell Thomas reported directly from the battle zones.

In 1941, the Commission was concerned about possible chain monopolies in broadcasting. The Commission, as reported in Broadcasting magazine in 1982, banned "option time, exclusive affiliations, ownership of more than one station in a market, or operation of more than one network by the same interests." The vote was 5-2, with

Commissioners Case and T. A. M. Craven dissenting. Case, in his dissent, wrote Broadcasting magazine, feared that "the proposals of the majority will result inevitably in impaired efficiency of the existing broadcast organization of the country."

Norman S. Case retired to his home state of Rhode Island and led a peaceful life with his wife, Emma. He died quietly in 1967 at the age of 79, survived by his wife and three children; Norman, John, and Elizabeth.

By Gerald V. Flannery, Ph.D. and James Nunez, M.S. candidate.

WALKER, PAUL A.
1934-1953

Paul A. Walker was born one of the eight children of Joseph Lewis and Hannah Jane Walker, in West Pike Run Township, Pennsylvania on January 11, 1881. On his father's side of the family Walker was of Virginia Quaker stock, and more remotely, Welsh. Walker received his preparatory education in the Southwestern State Normal School at California, Pennsylvania; upon graduation in 1899, he was class orator and on that occasion spoke on the evils of monopoly.

After normal school Walker entered the Chicago Institute which merged with the University of Chicago while he was still a student there. With an interest in dramatic activities and speech, he was awarded a quarterly scholarship in debate and the Ferdinand Peck prize in declamation. Walker also attended the John B. Stetson University in Florida for a while. In 1909, he finally obtained a Ph.B. degree from the University of Chicago, nine years after he first became a student at that institution.

The first major interruption in his studies occurred in 1904, when he became an athletic coach in a high school at Charleston, Illinois. The following year he moved to Shawnee, Oklahoma to become principal of its high school. During his tenure there, he introduced a system of student self-government.

In 1909, having left his high school position the year before, Walker entered the first class of the University of Oklahoma Law School. As the university debating coach he continued teaching during his law school days. Also, as chairman of a student's committee, at one time Walker helped persuade the legislature and governor to approve an appropriation for the law school building, and the dedication exercises. In 1914, two years after he had finished his law degree, he spoke for the alumni.

During the years that he was a debate coach, Walker met Myra Williams, a Latin teacher, whom he married in 1914. They both lived in Oklahoma City, where Walker became friends with Senator T. W. Elmer

Thomas and developed an interest in politics. In 1915 he began his long association with the State Corporation Committee of Oklahoma, which that year made him counsel. He held that position until 1919. In that post he stressed measures for conservation of oil and natural gas and helped draft legislation to further conservation generally. In 1919 he was appointed referee by the Supreme Court of Oklahoma under the statute that provided an assistant to the court to enable it to dispose of accumulated undecided cases. At the conclusion of his service with the court in 1921, he resumed his work for the State Corporation Commission, as special counsel. Among his duties was to conduct litigation before the commission at the State Commerce Commission.

As a Democratic candidate for membership on the Corporation Commission of Oklahoma in 1930, Walker won the office by popular vote, and the following year was chosen Chairman by the other commissioners. As Chairman he conducted a general investigation into the rates charged for natural gas, electricity, and telephone service in Oklahoma.

During this period he participated as well in a rate investigation by the Interstate Commerce Commission. During the investigation he sat as Chairman of the legal committee. A major problem he encountered as State Commissioner was the regulation of local utility companies which were wholly owned by corporations organized in other states. The controlling corporation often refused to produce its books for inspection on the grounds that the commission's jurisdiction did not extend beyond Oklahoma.

Walker created a report for the Corporation Commission, affirmed by the Supreme Court of Oklahoma in 1934, which formed the basis for rate reductions effected through the Federal Power Commission. Walker resigned from the Chairmanship of the National Association of Railroad and Utilities Commission in 1934, when he was called to Washington, D.C.

Before 1934, federal regulation of interstate communication was in the hands of the Interstate Commerce Commission and such bodies such as the Federal Radio Commission.

The Communications Act of 1934 created the Federal Communications Commission to provide unified handling of the problems related to radio, telegraph and telephone operations. Walker was one of the witnesses who testified in favor of the FCC at congressional hearings. He was also among the seven Commissioners appointed to the newly formed FCC. For a time, Walker held the record for the longest period with the FCC, until Rosel Hyde went on to amass a 41 year record.

During his almost twenty years of service, he saw radio and television come of age. He is best remembered as head of the division that conducted a three year investigation of A T & T. Walker, however, did not consider this his major accomplishment; he felt it was his work in allocating 242 television channels for educational purposes, having earlier pressed for educational radio.

When the FCC was formed, Walker was put in charge of the telephone division and an investigation of the telephone industry started soon after his appointment. The subsequent hearing produced 8,000 pages of testimony and over 2400 exhibits. The final report, signed by Walker in 1939, recommended rate decreases in long distance phone charges then, and in most of the succeeding years, until the end of World War II brought nationwide lower rates.

Broadcasting magazine, in its 1971 chronology of radio events, listed some of the items the FRC dealt with during his tenure. They were: the struggle between ASCAP and station owners represented by the NAB; that phonograph companies were labelling their records "Not For Use On Radio;" more American households had a radio than ever before, despite the depression; radio placed all its facilities at the disposal of President Franklin D. Roosevelt during the banking crisis; Associated Press members voted to ban network broadcasts of AP news; the FCC began scoring stations on their performance at renewal times; the invention and later introduction of FM broadcasting by Major E.H. Armstrong; the expansion of the Mutual Radio Network; the introduction of the A.C. Nielson metered rating system; the struggle of owners with the American Federation of Musicians and its threatened strike, the famous Mayflower Case; and problems of radio in wartime.

By Gerald V. Flannery, Ph.D. and Marisol Ochoa Konczal, M S.

PRALL, ANNING S.
1935-1937

Anning Prall was well known as an advocate of free speech. He reflected his views as both a congressman and a member of the Federal Communications Commission. Born in Port Richmond, Staten Island, New York, on September 17, 1869, he was the son of William Henry and Josephine Rebecca Prall. Anning Prall was educated in the public school system on Staten Island. He went on to attend New York University before accepting a position with the New York World newspaper.

In 1896, Prall left his position as a reporter to assume the position of chief clerk of the County Clerk's office. Between the years of 1907 and 1918, he functioned as a member of the head of the Mortgage and Loan Department of the Staten Island Savings Bank. During this time, he also served a brief term in the Department of Justice's Bureau of Investigation. Prall went on to become a three-time president of the New York Schools' Board of Education from 1919 to 1921. He resigned his post to become a Commissioner on the Board of Taxes and Assessments.

Prall focused his energy in directions other than public service. While the Commissioner of the Board of Taxes and Assessments, Prall was dealing in real estate, he was active in a number of social and civic organizations, and was the vice-president of the Jamestown Company, a coal mining operation in Pennsylvania.

In 1923, he was elected to the United States House of Representatives from the 11th District on Staten Island, Bedloe's Island, Governor's Island, and the northern tip of Manhattan. Prall served as a congressman for twelve years in the 68th through the 73rd Congresses. He served as a member of the House Committee on Banking and Currency during his term. Additionally, he was a personal friend of President Franklin D. Roosevelt, who would later name him to the Federal Communications Commission. Prall retired his membership from the House of Representatives on January 3, 1935.

Prall was chosen by his peers to become the Chairman of the

Broadcast Division soon after he assumed office. His term lasted only from January to March of 1935. At the end of his brief term, he was reappointed by President Roosevelt. Prall was named Chairman of the Commission and held the position until his death.

Under Chairman Prall, the FCC moved to a consumer-oriented activist stance that included warnings to stations about accepting medical advertising that claimed "cure-alls." The FCC reemphasized the need for the broadcast industry to adhere to the rules. In the wake of the Communications Act, Prall rallied the FCC behind his concern for good taste in broadcasting and organized a national meeting of network heads. The FCC was also trying to establish better relationships between radio and education as Prall recognized a great potential in combining the two.

In a move to flex its regulatory muscles, the FCC declared in October of 1935 that it would begin granting only temporary license renewals to stations whose programming was under investigation. The 1936 FCC required the announcements of recordings and transcriptions to a limit of one per quarter hour.

During his tenure, Prall saw the repeal of the Davis Amendment to the Communications Act. The amendment had required an equal division of radio stations in five geographic zones. The zones covered one or more states. Prall arranged a hearing on the new allocations of frequencies / power that led to the Supreme Court decision in 1938 that set the policy on multiple ownership for the next thirty years.

During his tenure as Chairman, Prall defended the freedom of speech over the airwaves, he encouraged the development of amateur radio operators, and did much to "clean house" of what he believed to be bad programming.

One of Prall's first actions in office was to send out a warning shot to the radio industry proclaiming that stations would be wise to regulate their activities voluntarily rather than subject themselves to legislative action. Prall realized the power of nonrenewal of licenses as leverage to influence the violators of FCC legislation.

Prall also realized the need to develop the educational, cultural, and public service potential of radio. So staunch was Prall in support of freedom of speech on radio, that he ordered a station to broadcast a speech by Communist presidential candidate, Carl Bowder, in 1936. Bowder had previously been denied that right.

Broadcasting magazine, in its 1971 chronology of radio events, listed some of the items the FRC dealt with during his tenure. They were: the FCC began scoring stations on their performance at renewal times; the invention and later introduction of FM broadcasting by Major E.H.

Armstrong; the expansion of the Mutual Radio Network; the introduction of the A.C. Nielsen metered rating system; the struggle of owners with the American Federation of Musicians and its threatened strike.

    Prall's tenure of service ended with his death January 23, 1937 at Boothbay Harbor, Maine. He was survived by his wife, Jane, and two sons, Anning and Bryan.

By Gerald V. Flannery, Ph.D. and David A. Male, M.S.

CRAVEN, TUNIS A.M.
1937-1944, 1956-1963

Tunis Augustus MacDonough Craven, an Episcopalian from Philadelphia, Pennsylvania was twice appointed to the Federal Communications Commission. He was the first and only person to receive two appointments. His first term ran from August 21, 1937 to June 30, 1944 and the second from July 2, 1956 to January 31, 1963.

He was the son of T.A. and Harriet Baker (Austin) Craven. His prep school education was at St. Paul's school in Baltimore, Maryland from 1902-08. He graduated from the U.S.Naval Academy at Annapolis in 1913. In 1915 he married Josephine La Tourette and had two children. In December 1931 he married a second time to Emma Stoner and had another child. He later married Margaret Preston and in March 1971 he remarried Emma Stoner.

After graduating from the Naval Academy, he was assigned as a Radio Officer on the U.S.S. Delaware from 1913 until 1915 just before World War I, during the Mexican campaign. He specialized in radio communications while he served in the navy and on the battleship U.S. Asiatic, he was a Fleet Radio Officer. He also controlled naval ship-to-shore and transoceanic radio services. Craven was in charge of the development, manufacture, and the purchase of naval ship radio installations and also became the administrative director of radio research and design for the navy. He was the first to modernize the fleet radio communications system and was commended for developing a way of transmitting orders without giving away the location of ships in submarine-infested waters during World War I.

Some of his accomplishments included: U.S. Naval Representative at Provisional Inter-allied Communications Conference at Paris, France, 1919; and Naval Representative on U.S. Government Inter-Departmental Board to arrange for collection and distribution of meteorological data, 1919. Other organizations were the U. S. Naval Institute, Ends of the Earth, Military Order of the Loyal Legion of the U.S., Army, Navy, and Marine Corps Country Club, and Kilocycle-Wave Length Club.

After the war he served in the capacity of Chairman of sub-committee on wave-length allocation at a International Conference in Washington, and U.S. Naval Radio Technical Advisor at a International Conference on electrical communications. In 1921 he was a U. S. Naval Representative at a Conference of the Radio Technical Committee on International Radio Communication in Paris and was also Battleship Force Radio Officer during that year. From 1921-22 he was a Fleet Radio Officer of the U.S. Atlantic Fleet, next he was U.S. Fleet Officer from 1922 to 1923 and then he was made director from 1923 to 1926 of the Bureau of Engineering.

Later, he was a member of Inter-Departmental Radio Advisory Committee and a Radio Technical Advisor for Radio frequency allocation at an International Radio Conference in Washington, D.C. The secretary of Navy presented Craven with a letter of commendation for work at the International Radio Conference in Washington, D.C. in 1927. In 1930 he resigned from the Navy to work in private radio as a consultant engineer, although he served as Commander in the U.S. Naval Reserve until 1944. From 1930 to 1935 he was vice-president in technical matters for many broadcasting companies. He is credited with developing directional antennas in order to expand broadcasting.

He also did some of his work as an author. He made some notable contributions in his writings to the development of radio. Along with Captain Hooper he wrote Robinson's Manual of Radio Telegraphy and Radio Telephony. He had many other technical essays on naval radio communications to his credit.

Craven was named as Chief Engineer of the Federal Communications Commission in 1935 and he held that position until he was appointed as Commissioner on August 25, 1935. His jump to Commissioner, according to The New York Times, was part of a reorganization of the FCC and seen as a way to help bring the affairs of the FCC in order. Because of Craven's twenty years of radio experience, he was seen as the one for that job. Yet his appointment was not one that pleased FCC Commissioner George Henry Payne who was quite often in disagreement with Chief Engineer Craven. Congress engaged in a heated discussion about Craven's qualifications and the FCC was chastised for his appointment. However, Craven's qualifications could not be disputed with his broad experience in radio.

On June 30, 1944 Craven's first term as a FCC Commissioner ended when he requested President Franklin D. Roosevelt not reappoint him for another term. He said that he planned to become associated with a publishing and radio station licensing interest in the Midwest. Cowles Broadcasting Company made him vice-president in charge of technical areas and in that capacity he worked until 1949. During this time he was also a member of the board of directors of the National Association of Broadcasters.

Craven worked with the firm of Craven, Lohnos and Culver, as a consulting engineer from 1949 until 1956. He was a Fellow of the Institute of Radio Engineers and former president of the Association of FCC Consulting Engineers. President Dwight D. Eisenhower appointed Craven for a seven year term to succeed Edward M. Webster; on July 2, 1956 he again became a FCC Commissioner and made history by being the only Commissioner to serve two non-consecutive terms.

The FCC was again being charged with inefficient operation and was believed to have been involved in unethical behavior: it was being accused of scandalous management. The FCC was in a similar condition when Craven was first appointed, yet this time there were no questions about his qualifications.

Broadcasting magazine, in its 1971 chronology of radio events, listed some of the items the FRC dealt with during his tenure. They were: the struggle between ASCAP and station owners represented by the NAB, and that phonograph companies were labeling their records "Not For Use On Radio." More American households had a radio than ever before, despite the depression; radio placed all its facilities at the disposal of President Franklin Roosevelt during the banking crisis; Associated Press members voted to ban network broadcasts of AP news; the FCC began scoring stations on their performance at renewal time;. the invention and later introduction of FM broadcasting by Major E.H. Armstrong;, the expansion of the Mutual Radio Network; the introduction of the A.C. Nielsen metered rating system; the struggle of owners with the American Federation of Musicians and its threatened strike; the famous Mayflower Case;, and problems of radio in wartime.

In 1956, after 50 years of government service, Craven stepped down after a wonderfully distinguished record due to mandatory retirement. He was replaced by Kenneth A. Cox, Chief of the FCC's Broadcasting Bureau. Craven continued to serve the FCC and the government as a consultant and in October, 1963 he was one of the key representatives at the Geneva Conference on space communications allocations.

A special ruling given by the comptroller general was necessary for Craven to be allowed to act as consultant to the Commission because of his age. Chairman Newton Minow stated that the FCC was extremely grateful for Craven's counsel and guidance and his talents would be an important asset to the United States. He was given an office in the FCC headquarters and worked closely with legal assistant Robert Koteen and engineering assistant Fred Heister on space and telecommunications matters, and satellite communications.

Broadcasting magazine for December 17, 1962 quoted President John F. Kennedy in a letter to the FCC's space Commissioner:

> ....All of us in this administration place the highest priority on the program already well-advanced, to bring into being at the earliest practicable date an operational global satellite communications system...The fact that you are willing to undertake this added assignment after concluding over 36 years of distinguished federal service is indeed a tribute to your devotion to duty.

According to Les Brown's New York Times Encyclopedia of Television, Craven:

> opposed any proposals by the Commission that smacked of government control over programming or over business practices, and he held fast to the view that decisions on programming in the public interest were best left to the judgment of broadcasters.

Praised by many, the veteran engineer, Commissioner T.A.M. Craven gave many years of hard work and devotion to the FCC. On May 31, 1972, just nine years after he left his position as Commissioner, he died in his hometown McLean, Virginia.

By Brenda Trahan, M.S .and Gerald V. Flannery, Ph.D.

McNINCH, FRANK
1937-1939

President Franklin D. Roosevelt borrowed Frank McNinch from the Federal Power Commission in August 1937 to place him on the Federal Communications Commission. The FCC and the broadcast industry were being charged with trafficking in licenses, political influence and lobbying, and information leaks due to dissension among Commissioners. McNinch was the man the President selected to put the affairs of the FCC in order. McNinch never had any problems when in came to putting things in order. As a young mayor and commissioner of finance in Charlotte, N.C. he stirred up considerable local criticism when he deputized citizens as special police to restore order in the streets due to violence resulting from a street car strike. This action did generate considerable local criticism but he was re-elected in 1919 and served out his Democratic term. He later won re-election by a margin of three-to-one.

During the presidential campaign of 1928, his local effectiveness as a political power received national attention. McNinch had remained on the Democratic ticket after the race for mayor in Charlotte, despite being considered by many as a staunch prohibitionist. The democratic party supported the "wet" candidate, Alfred E. Smith, but McNinch organized and became Chairman of the anti-Smith Democrats, who were able to swing North Carolina into the Herbert Hoover column. It was the first time in the history of the state that the Republican candidate got a majority.

On January 1, 1931, President Hoover appointed McNinch to the Federal Power Commission. At first McNinch declined the post, saying that he had never given any attention to utility problems, and that most of North Carolina's Democrats demanded McNinch not accept the position because he had proven his loyalties lay with Hoover during the presidential election. The president renewed the offer and McNinch accepted.

McNinch was born in Charlotte, N.C., April 27, 1873, son of Franklin

Alonzo and Sarah Virginia. He was of Irish decent. His father owned and operated a hotel. McNinch was educated in the public schools in Charlotte for his preliminary education, and continued at Barrier's Institute in that city. He graduated with an LL.B at the University of North Carolina in 1899. The following year he began his legal practice in the firm of McNinch and Kirkpatrick. He was a member of the North Carolina House Of Representatives in 1905 and was elected mayor of Charlotte in 1917.

McNinch remained a power-hitter taking control of the FCC in 1933 when Roosevelt designated him Chairman, a position he held until 1937. Being Chairman well suited McNinch for he had attended the world Power Conference at The Hague, Netherlands, in 1935. Later, he became vice-chairman of the National Power Policy Committee and a member of the National Emergency Council.

As Chairman of the Power Commission his authority was restricted to water power projects on navigable streams over which Congress had jurisdiction and the power to set rates, control services, and security issues. After constant agitation, Congress passed the Federal Power Act of 1935 which gave the Commission jurisdiction over the interstate transmission and sale of electric energy, supervision of rates, issuances of securities, installation of uniform accounting systems and investigation into subjects concerned with the organization, management and operation of electric public utilities. Still more power was added to the Commission when its duties were extended to include administration of the Flood Control Act of 1938 and the Natural Gas Act. During his reign on the FPC, McNinch changed its way of doing business. He successfully broke up many powerful groups that were dominating the utility industry.

McNinch is given much credit for creating the "death sentence" clause in the Public Utility Act of 1935. This act greatly dampened the unbridled power of certain financial groups who had been controlling the electric power market for years. McNinch, at this point in his career was respected by many for his ability to lead effectively, and hated by others who found themselves on the street after he cleaned house.

During his administration as Chairman of the FCC, Commission procedure was reorganized, many changes made in its personnel, its work speeded up. Once again McNinch began to work on the "big guys' in order to save the "little guys." McNinch was known as a hard-nosed Commissioner who wasn't afraid to investigate anything that might lead to fairness in communications. He was the first to initiate investigations of possible monopolies in radio networks and local station contracts and relations. In September of 1939, he resigned from the Commission and became special assistant to the attorney-general of the United States. He continued in that capacity until 1946 when he retired from public life.

McNinch was a Mason and Knight of Pythias. He was a Life-Elder and Bible-teacher at the New York Avenue Presbyterian Church in Washington, D.C. McNinch was married twice, widowed by his first wife Mary and left with three children; he married his first wife's sister and had two more children from that marriage. Frank Ramsay McNinch died April 2, 1950. He had been in ill health for some time and developed pneumonia.

By Gerald V. Flannery, Ph.D. and Ricky L. Jobe, M.S.

THOMPSON, FREDERICK I.
1939-1941

Frederick Ingate Thompson, president and publisher of the Montgomery Alabama Journal and Times, was named by President Franklin D. Roosevelt to the Federal Communications Commission on April 8, 1939. He succeeded Judge Eugene O. Sykes, who had been Chairman of both the Federal Radio Commission and the FCC. Thompson was then 63 years old so the president knew he would not serve long.

Thompson was born in Aberdeen, Mississippi on September 29, 1875 to Laura Cox and Edward Paul Thompson. He had an early start in news writing, being only seventeen years old when he became editor of the Aberdeen weekly, and from that point on he was always closely identified with the newspaper business. He left the Aberdeen paper in 1875, after three years, to accept the editorship of the weekly Commercial Appeal in Memphis. It was there that he met Adrianna Ingate of Mobile, Alabama. They were married in the Episcopal Church on February 5, 1900. After five years he left the Commercial Appeal to join the firm of Smith and Thompson, news-paper representatives with offices in New York and Chicago. He left that firm in 1908, after nearly seven years of service. One year later he became chief owner and publisher of the Mobile Daily and Sunday Register. He stayed there until 1932. During this time, from 1916 to 1932, he owned / published the Mobile News Item, an evening publication. From 1922 to 1927 he also was the owner / publisher of the Birmingham Alabama Daily, another morning paper, and the Sunday Age-Herald. He was also a Director of the Associated Press for nearly ten years.

Thompson was a delegate to the Democratic National Convention in 1912, 1924 and 1928. He was appointed to the Alabama Educational Commission in 1919; then in 1920 he was appointed Commissioner of the United States Shipping Board by President Woodrow Wilson. Thompson was reappointed by both President Warren Harding and President Calvin Coolidge. He resigned from the Shipping Board in 1925.

President Franklin D. Roosevelt appointed Thompson to the Advisory Board on Public Works in 1933; then in 1935 he became a member of the Alabama State Docks Commission. He had also served as president of the Seaboard Investment Company. Thompson, joined the FCC on April 8, 1939 and left on June 30, 1941, a few months before Pearl Harbor. The war would bring about many changes in America and in communications from rationing to voluntary censorship. It would delay the nationwide development of television until the late forties and radio would become a premier news source about the war from overseas with front line reports from correspondents to widespread commentary by a few of the premier reporters such as Edward R. Murrow.

Broadcasting magazine, in its 1971 chronology of events, listed a number of things the FCC dealt with in that period. They were: the telecast of the opening of the 1939 World's Fair; the lifting of the FCC ban on international broadcasts; the voluntary ban on liquor advertising by the National Association of Broadcasters; the lengthening of radio licenses; the consideration of economic factors in the award of any license; the organization of the Transcontinental Broadcast System by Elliot Roosevelt; the introduction of triple FM relay broadcasting; the authorization of 35 channels for FM radio; the investigation of "Pot Of Gold" and other giveaway programs on radio; the FCC approval of "limited commercialization" on television; broadcasters won the right to put records on the air without getting permission from everyone who took part in the creation of the record; the creation of the American Broadcasting Company; the hearing on radio ownership by newspaper companies; and the FCC fixed the standard for television at 525 lines.

Thompson died on February 20, 1952 in a Mobile hospital.

By Gerald V. Flannery, Ph.D. and Bobbie DeCuir, M.S,

FLY, JAMES L.
1939-1944

James Lawrence Fly, a lanky, six-feet three-inch, begoggled Texan with sandy "moth-eaten" hair (his own adjective), and blue eyes, was described by some as friendly, naive, or just a big country boy. More often he was said to be arrogant, offensive, hot-tempered, unfair, even ruthless, and to have a Southwesterner's love of a bang-up fight. A man of contradictions, Fly was nominated as a member of the Federal Communications Commission by President Franklin D. Roosevelt on July 27, 1939. Senate confirmation without opposition came on August 1 and the oath of office was taken by Fly on September 10. At that time he was appointed Chairman of the FCC succeeding Frank R. McNinch. The Commission Fly inherited was seen as weak and with no settled over-all policy. The members were described as being at swords' point with each other. Under Fly's strong management and direction, the Commission established a commanding place for itself. According to some, his leadership was so strong that Fly was not merely the Chairman, he was the Commission.

The son of Joseph Lawrence and Jane Ard Fly, James Lawrence Fly was born on February 22, 1898 in Seagoville, Dallas County, Texas, where his family had resided since the Civil War. Two grandfathers fought for the Confederacy, one died. His roots can be traced back to Jamestown, 1636, with a tradition of politically minded ancestors who were persuasive and hot-tempered, but above all, competent. His brother's were in politics and Fly spent much of his boyhood learning from the school of practical politics, the courthouse steps, just what was done in the political arena, to whom it was done, and how to sidestep it.

Fly graduated from Dallas High School in 1916 and received an appointment to Annapolis where he distinguished himself, graduating in 1920 with various prizes and commendations. While in the Navy, he became a kind of lawyer who was responsible for prosecuting or representing officers up for naval discipline. Fly served aboard the U.S.S. Idaho and the U.S.S. Baltimore before retiring in 1923 after an accident which caused superficial damage to the nerves of his scalp resulting in the loss of hair in patches, hence, the self-ascribed adjective "moth-eaten" hair. That same year he married Mildred Marvin Jones by whom he had two children: James and Sara.

In 1926, Fly received his L.L.B. from Harvard Law School and joined the firm of White and Case in New York City. From 1929 to 1934 he served Thurman Arnold as Special Assistant to the U.S. Attorney General during which time he was in charge of antitrust prosecutions. As head of the legal department of the Tennessee Valley Authority he came to grips with the powerful and prominent Wendell Willkie, then president of the Commonwealth and Southern Utilities which was based in New York City. When the battle was over, Wilkie was quoted as having said, "He is the most dangerous man in America, to have on the other side."

As Chairman of the Commission, Fly turned the once feeble Commission into a strong regulatory agency, backed by the President and the courts, but the battles were long and dirty. Among his accomplishments were the improved efficiency of the U.S. telegraph service, through a merger of the Postal and Western Union, and a reduction of long-distance rates, which saved the public nearly a quarter of a billion dollars a year. On September 15, 1939, the networks, in an unprecedented self-regulating move, drafted a code for war coverage which stressed full, factual reporting with a minimum of sensationalism, suspense or undue excitement, something which had occurred at the beginning of the European conflict. Although there was fear that the Commission was on the verge of stepping in to take charge of the situation, Chairman Fly said the Commission had no plan to control war news broadcasts.

By February 29th, 1940 the FCC announced that RCA and smaller telecasters could proceed with limited telecasting beginning September 1. RCA began advertising their television receivers ($395) and started a drive to put a minimum of 25,000 TV sets in homes in the New York metropolitan area. On April 1, the FCC accused RCA of trying to put a "freeze" on television transmission standards and suspended its permission to transmit. Fly's critics called this move a "usurpation of power." Fly defended his position by saying that the television set should not be forced on the public when no standards has been set and a receiver which was useful today might be obsolete tomorrow.

In June 1940, the FCC Chain Monopoly Committee recommended drastic changes in network operations, such as limiting network ownership of stations and length of affiliation contracts. Fly felt that the broadcast industry had little idea of its public responsibility. He also felt it was very dangerous to have a "duopoly" dominated by the two national networks (NBC and CBS) and Fly set out to break it.

On March 31, 1941 a group of about 100 newspaper publishers named Mark Ethridge of the <u>Louisville Courier-Journal</u> to chair a steering committee which was to investigate the proposed governmental action outlawing newspaper ownership of broadcast stations. Ethridge was to survey the entire broadcasting situation for President Franklin D. Roosevelt, but before he could begin Fly published a report on radio as a monopoly. The name calling began and Fly responded by borrowing a line from John Randolph which likened the broadcasters' association to "A dead mackerel in the moonlight - it both shines and stinks."

On May 5, the FCC authorized full commercial operation for TV, fixing standards at 525 lines, 30 frames, and FM sound. At this time a major reorganization of radio network operations was called for by the FCC which would ban option time, territorial exclusivity, ownership of more than one station in a market, long-term affiliation contracts, and other things.

Angered, the industry fought back by supporting a plan for the Senate to investigate the FCC. They asked for legislation to aid the broadcasting industry. In the end the Supreme Court upheld the FCC's right to regulate broadcasting practices, specifically the right to compel the network to comply with the monopoly rule.

The House approved a resolution to investigate the FCC and Representative Eugene Cox was appointed Chairman of the committee. Fly was able to discredit Cox and the entire Congressional investigation did little more than prove Fly's competence.

On November 6, 1944 Fly resigned from the FCC to open his own law practice in New York. By the time of his departure, the top executives of the major U.S. networks and independent U.S. radio operators who once thought Fly wanted to control the content of their programs had come to the realization that Fly was instead their bulwark against Government ownership. These same executives begged him to stay. All this was to no avail.

On January 6, 1966, James Lawrence Fly succumbed to cancer. He was 68.

By Fran DeLaun, M.S. candidate and Gerald V. Flannery, Ph.D.

WAKEFIELD, RAY C.
1941-1947

Ray Cecil Wakefield was born in Fresno, California on August 12, 1895. He received a B.A. degree in 1916 from Stanford University. He later received a J.D. in law from Stanford in 1918. He was admitted into the California State Bar Association in Fresno and was later appointed Deputy District Attorney in 1920. He worked for various law firms for 10 years, during which time he became Chairman of the Republican Central Committee in 1922. In 1928, he started his own practice and became the senior partner at Wakefield and Hansen in Fresno. He married Laureda Thompson in 1930. In 1932, he was voted a delegate to the Republican Convention in Chicago. That year also saw an end to his partnership in Wakefield and Hansen. In 1935 he began a new firm called Wakefield and Stanford. At this same time he became Inheritance Tax Appraiser for the state of California in Fresno until 1937. Also in 1937, Wakefield gave up his senior partnership in that firm and began his term with the California Railroad Commission, lasting from 1937-1941. He became president of the California Railroad Commission in 1938. He became a member of the National Association of Railroads and Utilities in 1939. Wakefield was an important force with the NARU Commission; as time passed he was elected second vice president. Following a successful tenure as president, Wakefield was voted to the executive committee. From 1937-1938 he was a member of the Interstate Commerce Commission at which time he took part of the freight rate increase case under investigation. Wakefield's political career seemed to flourish during the 40's. It was during this period that he was appointed to the Federal Communications Commission by President Franklin D. Roosevelt in 1941. He was nominated to the FCC to succeed the late Thad H. Brown and was sworn in on March 22, 1941. The FCC was not new to Wakefield because in 1938 he assisted with inquiries of the Pacific Telephone and Telegraph Corporation's telephone rates.

Wakefield was heavily involved in the debate that led he FCC put on the books what is called the AVCO ruling that said a radio station must advertise for bids when selling the station instead of conducting a private deal. It would have to auction the license.

Wakefield served as chairman or the Commission' s Telegraph Committee from 1941-1947. He later represented the United States in South America when communication problems began to occur in 1944-1945. He served as member to the U.S. delegation to the Third Inter-American TeleCommunications conference in Rio de Janeiro in 1945. Wakefield was a man who fought for the public. He wanted to save the public money in their use of radio and cable.

The nation was coming out of the depression when Wakefield joined the FCC but all eyes were on Europe where Germany was using radio as a propaganda tool and was frighteningly effective with it. Adolph Hitler believed that initially Germany should invade a country by radio, newspapers and magazines, first from outside the country, then from within. One of his techniques was to use radio to point out the problems that were occurring in the country he was about to invade, then follow this with information about how well Germany was handling a similar area or problem. Once his forces invaded a country, he took over the mediums of mass communications, particularly radio, trying to keep everyone on staff, if they would present the German point of view; later, if the broadcasters would not willingly cooperate, Germany would gradually change the management and makeup of the station. Hitler's effectiveness led American scholars to decide that radio had a major effect on public opinion, a theory that would remain in currency for quite some time.

Broadcasting magazine, in its 1971 chronology of important events, listed a number of things the FCC dealt with in that time. They were: the investigation of newspaper ownership of radio; the authorization of commercial television; the adoption of the 525 line system; the banning of multiple station ownership in the same area; extending the license term to two years; the FCC freeze on station construction; the House investigation of FCC practices; the increase in ownership rule allow one person/group to own up to five stations; the FCC ordered identification of sponsorship on broadcast material; and the FCC issued its report on public service responsibility which becomes known as the "Blue Book;" and RCA demonstrated color.

Wakefield's term ended in July 1947 after he was renominated by President Harry Truman for another six-year term. A Senate subcommittee began investigating Wakefield's political affiliations and Truman withdrew the nomination without explanation. The president instead appointed Commissioner Robert F. Jones. Wakefield was then appointed Chairman of the United States delegation to the International

Provisional Radio Frequency Board meeting in Geneva, Switzerland. He returned to Washington in 1949 because of insomnia and ill-health. Four months later he was found on the bathroom floor of his home in Palo Alto, California with his wrist deeply slashed.(September 20, 1949). An unsuccessful operation was performed and ten days later he died. The cause of death was listed as suicide.

Political organizations or agencies were just one part, although a major part of Wakefield's life, but he spent many hours working in community and civic organizations. He managed his law firms, participated in policy making activities on a number of boards and worked with charitable groups. He served as the Director of Community Chest, an organization that raised funds for the poor and needy. He was director of the Young Mens Christian Association (YMCA) in Fresno; a member of the First Christian Church in Fresno; he was also a Mason, a member of Coif, Delta Chi and Kiwanis International.

The FCC, on September 30, 1949, said this about Wakefield: "He was friendly and gracious in manner, devoted and unselfish in spirit. We mourn his loss as a friend, former colleague and an able public servant."

By Gerald V. Flannery, Ph.D. and Carla P. Coffman, M.S.

DURR, CLIFFORD J.
1941-1948

Federal Communications Commissioner, Clifford Judkins Durr was born to an aristocratic Alabama family in Montgomery, Alabama on March 2, 1899. He was the son of John Wesley and Lucy (Judkins) Durr. He received his A B. from the University of Alabama in 1919. His studies were interrupted by a brief period of service in the United States Army in 1918. After his graduation he went abroad as a Rhodes Scholar at Oxford University in England. He received his B. A. in Jurisprudence (1922). He was admitted to the Alabama Bar in 1923, and practiced with the firm of Rushton, Crenshaw, and Rushton, in Montgomery. He next joined the firm Fawsett, Smart, and Shea, in Milwaukee, Wisconsin until 1925 at which time he returned to Birmingham, and joined Martin, Turner, and McWhorter until 1933. Durr married Virginia Heard Foster on April 5,1926. She was Justice Hugo Black's sister-in-law.

In 1933, Durr entered government service as assistant general counsel of the Reconstruction Finance Corporation, which helped to recapitalize banks that had failed during the Depression. He stayed with the RFC until 1941. Also at this time (1941) he was general counsel and then director of the Defense Plant Corporation, an agency charged with putting American industry on a wartime footing. The years 1940-41 were active ones for Durr as he was also Vice President and director of Rubber Resources Corporation. On October 13, 1941, Durr was nominated by President Franklin D. Roosevelt for a seven year term as a member of the Federal Communications Commission. Even though he had no previous experience with broadcasting or communications, he took office on June 30, 1941 and served his full seven year term.

Three years into his appointment (1944), Durr was fighting quietly and steadily for the people's interests. A courageous and high principled man, Durr stood his ground on such topics as: preventing advertising control of radio, providing radio service for the one-third of the U. S.

outside the "daytime service area", preventing high prices for the purchase of radio stations (to help the little fellow get in), balancing presentation of controversial issues on the air, allowing labor and consumer groups to buy air time easily, developing more local public service programs, and providing for an adequate nationwide telephone and telegraph service to reach all citizens. Durr was for the people. He was one of the first national leaders to tell the public what FM radio could mean for community education because of the low cost involved.

On March 7, 1946 the FCC issued a pronouncement under the title of Public Service Responsibilities of Broadcast Licenses, which came to be known as the "Blue Book". The Blue Book outlined what the Commission thought to be important programming in the public interest. It was based on a memo expressing some of Durr's philosophy about the regulation of broadcasting, not a rule or regulation. It was taken quite seriously by the Commission and adhered to for several years.

In a FCC report on radio's behavior (March 8, 1946) he insisted that radio's business was the FCC's business. He dissented so often from his fellow Commissioners that it is was news when he voted yes. Durr's beliefs were the same as those of Herbert Hoover who once made the statement, "The ether is a public medium, and its use must be for public benefit. The use of radio channels is justified only if there is a public benefit ." During June of 1946, Durr successfully got the FCC to hold 90 FM channels open for a year, to give returning servicemen a chance to bid for them.

The year 1947 was a time when J. Edgar Hoover's FBI investigations of possible disloyalty by U.S. citizens were startling the country and its leaders. Hoover sent memos to the FCC urging them not to give out licenses to certain people because of their suspected contacts with possible Russian activists. Durr was appalled at Hoover's accusations based only on rumor and hear say and he publicly stated so. This maddened Hoover. Durr never backed down even after his fellow Commissioners turned their heads. He would not make judgements on non-factual gossip.

Durr put on the books what is called the AVCO rule. Under that rule, a radio station owner, about to sell his station had to advertise for bids instead of conducting a private deal. This was to give the public a chance. He put up a fight against big newspaper publishers encroachment upon radio. He also gave FM radio its second chance and put several FM stations on reserve for veterans still in the service. To his credit as well goes the slowing of big advertising companies buying large blocks of time on stations. At this time many broadcasters did not like Durr, but there were those that were for him.

The summer of 1948 brought the end of Clifford Durr's seven year term with the FCC. He declined reappointment to the agency by President Harry Truman and retired to private business in Alabama. When the President asked him why he declined the nomination he replied that he could not conscientiously administer the president's loyalty program. Durr was feted at a luncheon upon his retirement, but there was a sense of doom that afternoon. Few of his colleagues were with him in his struggle to balance the powers between the audience and the station owner. The people attending the luncheon praised him but many were relieved to see him go.

In 1950 an article in The New York Times told of Professor Louis F. Budenz's accusation (a former communist) that Durr played a major role in the infiltration of Red Communism through the airwaves to the U.S. James L. Fly, former Commissioner, was said to have helped him accomplish this Communist infiltration. When he returned to Montgomery, Alabama to private practice in 1954, Durr was for a time the only lawyer in town who would take on cases by blacks charging violations of civil rights. It was Durr who arranged for the bail of Rosa Parks, the black seamstress charged with violating the city's bus segregation ordinance. Lyndon B. Johnson wrote him a letter praising the values concerning rights and race Johnson had learned from Durr.

Durr was a Democrat and a Presbyterian. He was married for 50 years to the former Virginia Foster, who was also active in civil liberties. They were the parents of four daughters, Mrs. Walter A. Lyon of Harrisburg, Pa., Mrs. F. Sheldon Hackney of Princeton, N. J., Mrs. Virginia D. Parker of Washington and Mrs. Richard V. Colan of Birmingham, Ala. He had two sisters, a brother and nine grandchildren at that time. Clifford Judkins Durr died at the age of 76. A brief list of his honors would include:

University Lecturer: American British Universities and UCLA (theater arts); Service in U.S. Army; Rhodes Scholar; Recipient of Variety Award 1946; School of Broad-casters Award 1947; Page One Award, American Newspaper Guild 1948; Lasker Civil Liberties award, N. Y. Civil Liberties Union 1966; Life membership in Institution for Education by Radio; member Phi Beta Kappa; Member Sigma Alpha Epsilon; Blue Letter in Rugby at Oxford.

By Gerald V. Flannery, Ph.D. and Carla P. Coffman, M.S.

JETT, EWELL K.
1944-1947

Ewell K. Jett was considered one of the most knowledgeable people in the radio industry during its Golden Age. Commissioner Jett was born on March 20, 1893 in Baltimore, Maryland and attended both elementary and high school there. When Jett was eighteen years old, he joined the Navy and began his radio career as a telegraph operator and radioman on the battleships U.S.S. Utah, the U.S.S. Michigan, and the destroyer, U.S.S. Parker. From 1914 to 1916 he was assigned to the Arlington radio station in Washington, and during this period the Arlington station conducted historic experimental radio telephony tests, involving short wave transoceanic transmissions.

In 1915, Jett married Viola Ward while on shore duty. During World War I, Jett held a temporary Commission as Warrant Radio Officer on the flagship U.S.S. Seattle and the Battleship U.S.S. Georgia. He was commissioned ensign in 1919, and was Officer-in-Charge of the Navy's transatlantic radio control station until 1922. From 1923 to 1926, then holding the rank of Lieutenant, Jett was aide on the staff of Admirals Chase and Marvel. In 1929 Jett was loaned to the Engineering Department of the Federal Radio Commission. He was the assistant to Lieutenant Commander Carven, a short wave expert. When Carven returned to Navy research work in the Department of Engineering, Jett took over the short wave work for the FRC. He retired from the Navy in 1929 to become the FRC's senior radio engineer for nonbroadcast engineering.

The FRC in 1930 was discussing what future television might have in broadcasting sports like baseball and football. Jett was not as enthused as some of his fellow Commissioners; he thought television was in its experimental stage and wouldn't be much of an entertainment medium for some time to come. The following year he was promoted to Chief Engineer. From 1939 to 1941, Jett served as Chairman of the Interdepartment Radio Advisory Committee (IRAC) the group responsible for handling government frequency assignments.

In 1941, when nearly one thousand radio stations in the Western Hemisphere changed to new wavelengths, Jett was one of the experts who went on radio to explain why it was being done. World War II came along and Jett was appointed Chairman of the coordinating committee on the Board of War Communications. Four years later President Franklin D. Roosevelt named him to the Federal Communications Commission where he provided expertise in allocating broadcast frequencies. All told, he served fifteen years for the FRC and the FCC as one of their technical experts.

At the time of Jett's appointment the nation was beginning to think about a new industry, television, which has been demonstrated at the 1939 World's fair, but was being shelved by the war effort. Television standards were being established, nonetheless, with plentiful advice and criticism from lay people and experts alike. Jett suggested one way to clear the logjam might be to allow two systems to develop, one with the standards possible in wartime, another to come later with "vastly improved standards." FCC Chairman Fly rebuked that suggestion. Fly resigned the FCC in November 1944 and Roosevelt wrote a letter to Jett that said in part:

> I can well understand that with your other extensive duties you would not care to undertake the burden of the chairman's office permanently, however, I do hope you will take this work during the interim period." He served as Chairman for about six weeks until Paul A. Porter was appointed by the president.

Commissioner Jett was next elected Chairman (1946) of the United States delegation to the North American Regional Broadcasting Conference, and was a member of the U.S. delegation to the International Telecommunications Conference in Atlantic City in 1947, his last year with the FCC. In his letter of resignation, Jett said he wanted to get back to private life and devote himself to other work in the radio field. He was succeeded on the Commission by George E. Sterling.

Broadcasting magazine, in its 1971 chronology of important events, listed a number of things the FCC dealt with in that time. They were: the investigation of newspaper ownership of radio; the authorization of commercial television; the adoption of the 525 line system; the banning of multiple station ownership in the same area; extending the license term to two years; the FCC freeze on station construction; the House investigation of FCC practices; the increase in ownership rule allowing one person/group to own up to five stations; the FCC ordered identification of sponsorship on broadcast material; the FCC issued its report on public service responsibility which became known as the Blue Book; and RCA demonstrated its all-electronic system of color;

competing color systems were reviewed by the FCC.

After he left the FCC, Jett was picked to represent the U.S. on the International Provisional Frequency Board that met in Geneva in 1948. He became vice president of the Baltimore Sun newspapers and director of their radio interests, then helped establish WHAR-TV in Baltimore. He died on April 29, 1965 at the age of 72.

By Gerald V. Flannery, Ph.D. and Bobbie DeCuir, M.S.

PORTER, PAUL A.
1944-1946

Paul Aldermandt Porter was born October 6, 1904 in Joplin, Missouri. He was the son of John H. and Dolly Porter. The family moved to Winchester, Kentucky when he was very young. Porter worked as a reporter when he was 15 years old. The money was needed to help support his widowed mother and her eight children. After graduating from Kentucky Wesleyan College, Porter received his LL.B in 1927 from the University of Kentucky. While in college he became city editor of the Lexington Herald, and stayed with the paper until 1928, when he left to enter the private practice of law. Later Porter came back to the newspaper scene and held the position of general attorney for several newspaper chains. It was during this time that he began writing editorials. instead of running the "canned" editorials set up to be used in small papers. Porter wrote intense editorials that landed him his first government job. He worked under Secretary of Agriculture Henry A. Wallace as a legal assistant. In 1937, he left that position to become counsel at Columbia Broadcasting System, where he stayed until 1942.The war interrupted his career in 1940 and Porter became legal advisor and assistant to Chester C. Davies on the National Defense Advisory Commission. This Commission was formed to plan a possible food program for an America at war. Next, he became a deputy administrator of rent control at the OPA.

Porter became publicity director for the democrats in the 1944 presidential campaign, and when the election was won , he was named to a federal post near the end of 1944. James L. Fly had resigned as Chairman of the Federal Communications Commission before expiration of his seven-year term of office; and on December 21, 1944 President Franklin D. Roosevelt appointed him Chairman of the FCC.

Porter was chiefly occupied with the wartime operations of the Commission. His primary concern was policing the airways in connection with the national defense. In pursuit of this, the Radio Intelligence Division maintained surveillance to detect espionage or other illegal radio transmissions that might give away shipping secrets or damage the war effort.. Other activities included issuing licenses to radio stations and operators, and regulating interstate and foreign communications by telephone, telegraph ,cable, and radio. The FCC was also engaged in the promotion of safety at sea through the use of communication facilities.The responsibility which Porter inherited was the center of controversy among naval officers, the National Association of Broadcasters, the networks, and Congress.

In February 1945, Porter advised Director of War Mobilization and Reconversion James F. Byrnes that nineteen thousand miles of leased wire circuits, with seven hundred extensions, and a large number of telephone sets, previously used for the dissemination of racing information, had been made available for essential civilian use as a result of Brynes's ban on horse and dog racing the month before.

Next, a really complicated problem was presented to Porter when, late in March, Navy Secretary James Forrestal told Congress that the United States had no stated policy governing international communications, that anyone could set up a radio, telephone or cable company, and that this sort of free competition was out of date Forrestal advocated a monopolistic "chosen instrument" policy, that is, a merger sanctioned by Congress of all United States companies engaged in international communications into one big government-backed corporation so that the United States could compete with similar foreign companies and secretly send its diplomatic and commercial messages, Porter urged a consolidation of overseas services but did not believe the Forrestal idea feasible because the FCC would have the job of regulating a corporation that included five Cabinet members among its directors, from private business. Porter told the Senate Interstate Commerce Committee on International Communications that the taxpayer's stake was enormous. The total cost of combined plant and equipment of the thirteen companies involved in the planned merger at that time (1945) was in the neighborhood $143,500,000 and this took it out of the realm of possibility during wartime. Porter's point was made and the Forrestal dream ended. Porter was also effective in enabling FM broadcasting to come of age.

Porter joined the FCC while America was a war and one of the government's major concerns centered around espionage and how to monitor radio without censoring it. A simple thing like a weather report broadcast from an area that contained a naval base or a troop embarkation point could provide enemy subs and ships with data they could not get easily anywhere else. Federal radio monitors were afraid that some German or Japanese sympathizers were playing certain songs that were a signal that a deployment of ships was leaving a particular harbor, or that innocent comments by on-air people about what was happening in such areas would act as a tip off to enemy vessels.

Broadcasting magazine, in its 1971 chronology of important events, listed things the FCC dealt with in that time. They were: newspaper ownership of radio; the authorization of commercial television; the adoption of the 525 line system; the banning of multiple station ownership in the same area; extending the license term to two years; the FCC freeze on station construction; the House investigation of FCC practices; the increase in ownership rule allowing one person/group to own up to five stations; the FCC issued its report on public service responsibility which became known as the Blue Book, and RCA demonstrated its all-electronic system of color.

Porter left the FCC in 1948.

By Gerald V. Flannery, Ph.D. and Ricky L. Jobe, M.S.

HYDE, ROSEL H.
1946-1969

Rosel H. Hyde served the longest tenure of anyone on the Federal Communications Commission and was the only person to be appointed by both Democratic and Republican presidents. He served as Chairman of the FCC twice and was at its helm during some of its most difficult years. Hyde, a conservative, considered himself a liberal, partly because he preferred as little regulation as possible.

Broadcasting magazine called him the broadcaster's friend, a man whose philosophy as commissioner was simple: broadcasting will improve if the government climate is favorable. Ironically, on the same day he was being honored for 45 years in government employment -- 41 in the service of broadcasting -- he was cited for contempt by the Commerce Committee for refusing to turn over some FCC case records.

Hyde was born in Bannock County, Idaho on April 12, 1900. He received his education at the Utah Agricultural College in the early twenties, then went on to George Washington University in 1924 where he earned a law degree. His career with the government began in 1924 when he won a position with the Civil Service Commission, a job earned through a competitive Civil Service examination. He was on the staff of the Office of Public Buildings and Parks from 1925 to 1928 when he joined the Federal Radio Commission as an Assistant Attorney. As a member of the FRC staff, he participated in the first general frequency allocation proceedings in 1928 and was still on staff when the FRC became the FCC in 1934. President Harry Truman appointed him to the FCC in April 1946, calling it a merit promotion for outstanding service. Gradually, working his way up through the ranks, he wound up as general counsel for the FCC.

President Dwight D. Eisenhower named Hyde Chairman of the FCC in 1953, but in a departure from custom, made it for only one year. At the end of the year, when the president failed to act, the Commissioners named him Acting Chairman, a position he held from April 1954 until the following October when George C. McConnaughey was appointed chair. Hyde became Acting Chairman again in May 1966 when William

E. Henry resigned and was named permanent Chairman by President Lyndon Johnson the following month. Broadcasting magazine called it a move that signaled the end of the New Frontier era, an era marked by controversy and an era studded with proposals that would affect the foundations of the broadcasting industry.

Hyde's appointment came at a time when there were broad ideological conflicts among Commission members. Walter Emery, writing in his 1961 book Broadcasting and Government, said that intensifying the difficulty of the period was the "almost constant surveillance and intermeddling of Congress, aided and abetted by the critical clamor and outcries of the broadcasting industry, the cable operators, the trade press and other special interests, not to mention an aroused and sometimes hostile public." As a member of the FCC staff, Hyde participated in a lot of important business: the frequency allocation hearings in 1935; the network investigations of 1938; the opening up of FM and TV broadcasting in 1941; and the overall TV considerations between 1949 and 1952, big growth years in television; he also represented the U.S. in countless international telecommunications conferences.

Broadcasting magazine, writing in June 1966, summed it up this way:

> His track record...is good. During his first term as Chairman, in 1953- 54, he was instrumental in breaking the procedural bottleneck holding up the grants for television stations... after the four year freeze..., when the public was clamoring for service. With the agreement of competing parties, procedures were devised for making grants on the basis of paper pleadings. The hearing process... was sidetracked.

The Hyde years at the FCC were characterized by three major cases: 1) American Broadcasting Company / International Telephone and Telegraph (ABC / ITT); 2) WLBT, and 3) WHDH; plus the birth of Community Antenna Television (CATV), and an emphasis on educational television.

The question of CATV regulation had been under consideration by the FCC for several years when Chairman Hyde directed the Commission's attention to the problem of who was to regulate cable TV and how. His position became the majority view, namely that the public interest required the FCC regulate CATV, thus a number of rule-making proceedings resulted in regulation, despite serious opposition from the CATV Industry and congressional challenges about the FCC's right to regulate cable.

In the ABC / ITT case, which concerned the merger of two communication giants, a move ABC hoped would give it the financial depth that NBC and CBS had, Hyde voted with the majority to approve the marriage, but the Justice Department vetoed it. In the WLBT-TV (Jackson, Miss.) case, where a suit by the United Church of Christ established the right of citizen groups to have standing before the FCC, Hyde sided with the station, but the federal court, citing the public interest, sided with the citizens. Hyde also participated in the lengthy hearings of the WHDH (Boston, Mass.) case, in which some broadcasters felt community groups were getting too much encouragement to challenge license holders at renewal time; however, on the final vote, Hyde abstained.

A strong supporter of educational broadcasting, Hyde's interest in the educational opportunities of the media may have been a reflection of the view held by President Johnson, whose wife Lady Bird owned broadcast facilities. Johnson, a former teacher, was said to have envisioned a nation-wide network of TV stations available for education, particularly useful should a president want to talk, directly from his office, with school children, face to face, in their classrooms.

Hyde was a consistent dissenter to what was called the hard-line policy of his predecessors, Newton N. Minow and E. Williams Henry. As the first Republican to head the FCC during a Democratic administration, he was expected to provide a steady hand at the wheel of the FCC, and to achieve a record of practical accomplishments, rather than fight ideological battles. Broadcasters considered him a firm friend. When the FCC approved a proposal limiting network ownership to fifty per cent of prime time programs, he dissented; he dissented also when the Commission voted to limit single companies to owning three TV outlets in the top fifty markets.Two of his strong supporters on capitol hill were Republican Senate minority leader Everett M. Dirkson and Democratic Senator John Pastore, Chairman of the powerful Subcommittee on Communication. Hyde tried to retire in June 1969 after four consecutive terms, but President Richard Nixon got him to stay on until October when a new Chairman would be named. After 41 years with both the FRC and FCC, he became known as "the Dean of the FCC."

Portions of this article appeared in FEEDBACK, the Journal of the Broadcast Education Association.

By Gerald V. Flannery, Ph.D. and Edwin A. Meek, Ph.D.

DENNY, CHARLES R. JR.
1945-1947

Charles Ruthaven Denny Junior, the youngest man ever appointed to the Federal Communication Commission, was its youngest Chairman. Not only was he the youngest to serve on the FCC, but after leaving the Commission in 1947 he joined the National Broadcasting Company and soon became the youngest senior executive in radio. Denny was born in Baltimore, Md., on April 11, 1912. He attended public schools and graduated from Western High School in Washington, D.C. In 1933, he graduated with honors from Amherst College with his bachelor's degree and went on to Harvard Law School, earning his law degree in 1936. That year he joined the Washington, D.C. law firm of Covington, Burling, Rublee, Acheson & Shorb, and was with them for two years. From 1938-42, he was in the Lands Division of the U.S. Department of Justice, initially as an attorney in the Appellate Section, moving up to Assistant Chief, and finally as Chief of the section. He also was a special assistant to the U.S. Attorney General at the Department of Justice. Denny began work with the FCC in 1942, assuming the duties of assistant general counsel of administration and litigation. Within a year, he was named general counsel. As general counsel, Denny had the job of supervising the work of about 60 attorneys in the Law Department, representing the Commission before several Congressional committees, conducting many important radio and wire communications investigations and hearings, and serving as head of the Law Committee of the Board of War Communications.

The young attorney's first important Commission assignment was preparation of the Commission's legal brief on the question of its jurisdiction in promulgating the chain broadcasting regulations. Denny argued and won the case before the U.S. Supreme Court, which remanded the case to the U.S. District Court for the Southern District of New York. As general counsel, Denny also represented the Commission in hearings in April 1942 on the Sanders Bill to revise the Communications Act, and in the winter of 1943, he headed the FCC's presentation of hearings on the White-Wheeler Bill to revise the Communications Act. Denny also represented the FCC as counsel during two years of investigation and hearings by the Select Committee to Investigate the FCC in 1943-44.

His last assignment as general counsel before becoming an FCC commissioner was the handling of the important Commission hearings on post-war radio frequency allocations. In 25 days of hearing in the fall of 1944, 231 witnesses appeared and testified on technical and engineering aspects of all phases of radio activity. In becoming a Commission member, Denny, a Democrat, filled the vacancy created when the term of T.A.M. Craven expired on June 30, 1944. Denny was sworn in as the seventh member of the Commission on March 30, 1945, after being confirmed by the U.S. Senate on March 26, 1945. His term was to run until June 30, 1951.

Time magazine reported Denny's position on the FCC and radio this way:

> Our whole effort is to try to have an intelligent working relationship with the industry. In broadcasting, in particular, we are engaged in building two things of tremendous importance—FM and television, and if the job is to be done right, we've got to join together. There's no time for name calling and bickering. In the main, our objectives coincide....
> The FCC, by all its policies and thinking, points in the direction of the freest possible use of radio for the purpose of keeping the public fully informed on both sides of all questions. Radio definitely has that responsibility—just as it has thE responsibility for good entertainment.

Less than a year after assuming the Commissioner's post, Denny was named Acting Chairman by President Harry Truman to take the place of Paul A. Porter. On Dec. 4, 1946, Denny was made full Chairman. As such, Denny, described by Time magazine (December 16, 1946) as "high-strung and high-powered," took up the work Porter had initiated with the FCC's "Blue Book," which dictated that radio concern itself more with serving the public than serving advertisers.

As might be expected, reaction by the broadcast industry to the Blue Book was quite negative. In fact, opponents said the FCC Commissioners were "stooges for the Communists." Denny himself said supporters of the Blue Book had been called "obfuscators, intellectual smart-alecks, professional appeasers, guileful men, and astigmatic perverters of society." Denny, however, who had helped write the Blue

Book, stood his ground and helped set into motion the methods for enforcing the book's dictates.

About the Blue Book, Denny said at a convention of broadcasters in October 1946, "We will not bleach it." Under Denny, the Commission announced it would heavily consider public service when issuing and renewing broadcasting licenses. Newsweek magazine (July 19, 1948) summarized Denny's tenure this way:

> . . .one of the brightest of official Washington's younger generation, Denny was in the center of a jittery teeter-totter. On one side was the radio industry, smoldering over the government-control implications in the FCC's Blue Book, released in 1946. On the other side was Congress, viewing the Commission's exercise of power with a suspicion... But Denny sat tight and made more friends than enemies...

It was during Denny's Chairmanship that FM radio and television were coming into their own, which created new regulatory problems. Denny's forthrightness and desire to make improvements in the broadcast industry often put him at odds with members of Congress. He was above all, however, considered an effective FCC Chairman and able administrator. Denny left the FCC after only two years of a six-year term. In October 1947, he accepted an offer to become general counsel with the National Broadcasting Company at a salary of $35,000, which was three times that of his FCC compensation. Just as he did in government, Denny also moved quickly up the ranks at NBC. He soon was named executive vice president, replacing Frank Mullen, the man credited with helping make the company the nation's largest network.

Broadcasting magazine, in its 1971 chronology of important events, listed a number of things the FCC dealt with in that time. They were: the FCC ordered identification of sponsorship on broadcast materials; the FCC issued its report on public service responsibility which became known as the "Blue Book;" and RCA demonstrated color; the Associated Press offered to admit stations as associate members;   and ninety percent of American homes had a radio.

An Episcopalian, Denny and his wife, the former Elizabeth Woolsey, whom he married on Dec. 31, 1937, reared three children, Alison and Christine, and Charles II.

By Michael A. Konczal, M.S. and Gerald V.Flannery, Ph.D.

WILLS, WILLIAM H.
1945-1946

William Henry Wills, like many others, was appointed to the Federal Communications Commission based on his political successes rather than his knowledge of broadcasting. A two-term governor of Vermont, Wills was appointed to by President Harry Truman on June 13, 1945. He was confirmed by the Senate on July 12, approximately one month before the second world war came to a close with the nuclear bombing of Japan.

Wills was born on October 26, 1882 in Chicago, but moved to Vermont with his family and attended that state's public school system. He moved to Bennington in 1910 and began his own insurance and real estate business in 1915. Before that he had worked as a store clerk for fifteen years. He later served as director of the Vermont Mutual Fire Insurance Company, and then as director of the Bank of Bennington.

He met and married Hazel McLeod in 1914 and they had a daughter Anne, who eventually married Stanley Pike, a lieutenant with the U.S. Navy. Wills opened the Wills Insurance Agency in 1915 in Bennington, and incorporated the business in 1928, just one year before the stock market crash. His political ascendancy began just as the country was about to sink into the Great Depression.

Wills first success came when he was elected to the Vermont House of Representatives in 1929, and he followed that by being elected in 1931 and again in 1935 to the State Senate. He served as president pro-tem in the senate. When his senate term ended in 1937, he ran successfully for Lieutenant Governor of Vermont, staying in that capacity until 1941. He made it to the top of the heap and accomplished much during World War II when he served as Governor from 1941 to 1945.

As Vermont's wartime governor he was responsible for the creation of the post of industrial agent, an idea conceived to do two things: 1) stimulate the state economy during a period of sacrifice and shortages, and 2) create jobs to keep young people from leaving Vermont to seek employment elsewhere. He also helped develop legislative backing to improve the University of Vermont, and is credited with stream-lining government through weekly department meetings, plus creating a weekly radio program for the public called "Your State Government." He used the program to keep voters abreast on what was happening in government. Additionally, he was instrumental in establishing an insurance plan for state employees.

On January 13, 1944, Wills informed the press that he would not seek a third term as governor, citing health as the reason. As result, Senator George Aiken ran unopposed. Wills returned to the private sector in January 1945, resuming his insurance / real estate business started thirty years earlier. President Harry Truman named him to fill a opening on the Federal Communications Commission, a spot created by the departure of eleven year veteran Norman S. Case. It seemed ironic that Truman, a democrat, would appoint Wills, a republican, but the president was only following the 4-3 political rule in balancing the membership of the Commission. Wills, prior to the 1944 election, came out strongly for Republican Wendell Wilkie; speaking on the CBS network, he called Wilkie the only republican certain to beat the strongest democratic candidate.That didn't set well with other political friends.

In addition to maintaining a high political profile, Wills was an active member of his community, wherever he was. He served as secretary of the Bennington Savings and Loan Association and the American Institute of Property and Liability Underwriters, Inc. He was a Commissioner for real estate licenses for the state of Vermont, a trustee for the Vermont Soldier's home, and a trustee of the University of Vermont and Vermont Junior College. He was also president of the Board of Trustees of the Episcopalian Diocese of Vermont, corporator of the H.W. Putnam Hospital, director of the Goshen Camp for Crippled Children and a junior warden of St. Peter's Episcopal Church.

Wills was awarded an honorary LL.D degree from Norwich University and the University of Vermont in 1941. Middlebury College followed with an award in 1945. Other civic interests included membership in the Elks Club, the Bennington Business Men's Association and the Masonic Order.

Broadcasting magazine, in its 1971 chronology of events, listed things that happened during his brief term on the FCC. They were: the FCC issued its report on public service responsibility called the "Blue Book;" the Associated Press admitted broadcasters as associate members; and RCA demonstrated its all-electric system of color television.

His term as an FCC commissioner was rather short lived. He served for a total of 219 days. He died of a heart attack on March 6, 1946 in the Bryant Hotel in Brockton, Massachusetts, making his term among the briefest. His death came when the United States was on the threshold of economic greatness. Had he lived past the age of 63, he could have been a factor in the regulation of the host of communications advances that were a product of World War II.

By Gerald V. Flannery, Ph.D. and Joe Lynch, M.S.

WEBSTER, EDWARD M.
1947-1956

Edward Mount Webster was born in Washington, D.C. on February 28, 1889. His father worked as a U.S Treasury Department civil servant. He attended local public schools in Washington, then he entered the United States Coast Guard Academy, graduating in 1912. Webster's thirty years of governmental service were spent in the Coast Guard where many of his ideas and innovations are still in use today. Webster himself once commented that his greatest sense of accomplishment was the job of establishing the United States Coast Guard's famed "ship-to-shore" system of radiotelephone communications during the prohibition era of the "twenties." Webster served in the Coast Guard until he retired in 1923. Because of his expertise in international and domestic communications, Webster was recalled to active duty in the Coast Guard and served until 1934. He was again recalled during World War II and served as the Coast Guard's Chief Communications Officer during those tours of duty.

Webster did not let retirement from the Coast Guard slow down his personal service to the government. His friendship and acquaintance with Federal Communication Commission engineers brought him into close contact with the FCC, thus, in 1934, he became an FCC engineer himself. A year later he became Assistant Chief Engineer, responsible for supervision of radio and wire services, matters relating to radio operators, and operations of marine, aviation, experimental, emergency, and amateur radio.

During his eight years of service with the FCC, Webster was also instrumental in creating and building an effective organization within the agency. Because of his background and training, he was primarily concerned with marine, safety, and other categories labeled as "special services."

An original member of the Interdepartment Radio Advisory Committee, Webster assisted the State Department in preparing twenty-one international communications conferences. At four of these conferences, he served as the U.S. delegate chief, and most of his appointments were directly made by the President or the Secretary of State. Then in 1947, Webster became a Federal Communications Commission member. President Harry Truman nominated Webster to fill the unexpired term of Paul A. Porter. In July of 1949, Webster began his full seven-year term after winning reappointment to the Commission. While serving almost a decade on the FCC, he listened to, judged, voted, and ruled on thousands of cases. Many cases were milestones in the early developmental stages of television .

Webster was one of five Commissioners that ruled in favor of Columbia Broadcasting System's initial color program introduced in the late forties. CBS won its color battle when the FCC approved its color standards after several years of legal battles. There were numerous violent reactions from other networks and professionals in the communications business when the FCC handed down its "Second Report" approving CBS's standards. This case was just one of many along the rocky road traveled by Webster during his career as an FCC Commissioner. However, Webster was never known as a dissenter as he was in the majority in most of the cases he heard. He did take a definitive stand against subscription television. He also was a part of the FCC rulings on AM and FM station studio locations. In 1953, the FCC approved the merger of ABC and United Paramount Theaters in a 5-2 decision. Webster dissented, asking for a further study of UPT's qualifications.

In 1951, the FCC took a large step, initiating a plan which reserved ten percent of channel grants to educational non-commercial operations, and withheld them indefinitely from commercial use. Webster, in a partial dissent to the plan, did not agree; but he was up against a tough fighter, though, in Commissioner Freida Hennock who championed the cause of the educational non-commercial operators. She stirred up enough educators and mothers to get the educational reservations.

Broadcasting magazine, in its 1971 chronology of events, lists a number of things that happened during his tenure. They were: the FCC took a strong stand on any station censoring political broadcasts; it gave stations the right to editorialize; the National Association of Broadcasters developed a new voluntary code for its members; the FCC froze TV license awards then later lifted it with a new allocation plan; the FCC repealed the AVCO rule; some networks pulled out of the NAB in a temporary dispute; later ownership rules changed so that one person/group could own seven TV, seven AM and seven FM radio stations; transistor radios were introduced; the merger of ABC and United

Paramount Theaters was approved; pay-per-view television was attempted; a number of UHF stations went on the air; and license renewals were extended to three years.

Edward Mount Webster retired from the FCC in 1956. He died in his home twenty years later, in 1976 at the age of 81, surrounded by his wife and two children.

By Gerald V. Flannery, Ph.D. and James Nunez, M.S. candidate.

JONES, ROBERT F.
1947-1952

Robert Franklin Jones was born In Cairo, Ohio on June 25, 1907. He was the son of Jenkin Charles and Josephine Devine Jones. Jones was interested in politics from when he was a little boy, an interest that possibly stemmed from his father's passion for political affairs. Following his graduation from high school, Jones entered Ohio Northern University of Law, where he was a class orator. After receiving his law degree in 1929, and being admitted to the Ohio bar that same year, Jones started practicing in Lima, Ohio. In 1935 he began his long tenure of public service in the capacity of prosecuting attorney for Allen County. As a result of his success in that post he was nominated and elected to the 76th Congress. He also received the Republican nomination for Representative from the fourth Ohio district in 1933 and was elected to the 30th Congress that same year. Jones served in Congress for eight and-a-half years until he accepted an appointment to the Federal Communications Commission in 1947. Jones was considered an efficient congressman. During his first year in office he was a member of the House Appropriations Committee. He later served as Chairman of that committee's subcommittee on the activities of the Department of the Interior. He became a specialist in Administrative Law and also familiarized himself with common issues dealing with government regulation. He was always an advocate of economy in government.

When Jones was appointed to the FCC by President Harry Truman 1947, he was a minority stock holder of an Ohio Broadcasting concern. He had to give up his stock to accept his new position on the Commission. A Republican, Jones was appointed to the FCC simultaneously with the withdrawal of the nomination of Ray C. Wakefield.

On June 23, 1947, ten days after his new appointment, Jones faced charges by newspaper columnist Drew Pearson that he was intolerant of Catholic and Jewish believers. Pearson appeared before a Senate

Commerce subcommittee to assert that Jones was a former member of the Black Legion and that his father was a member of the Ku Klux Klan. Pearson went on to say that Jones' father took him to Klan meetings as a boy and introduced him as the youngest Klan member. On that same day Jones testified and denied all charges brought against him.The Senate confirmed his nomination to the FCC on July 11, 1947, and on July 12, he was sworn in. His appointment by President Truman came as a surprise because Jones has been a strong critic of President Franklin D. Roosevelt.

During his term as Commissioner, Jones was a controversial figure and often clashed with other Commission members and with Harry Plotnik, the counsel to the FCC. In the early 1950's, Jones was the target of many attacks from the networks because he soon established himself as a sharp critic of the network establishment . On September 19, 1952 Jones resigned from FCC to enter law practice in Washing-ton, D.C. with his friend Arthur Scharfeld.

Despite his resignation Jones did not remove himself from administrative law. Two years later, exactly on, August 5, 1954 Jones was named general counsel for an investigation of radio and television networks. The study was instigated by the Chairman of the Senate Interstate and Foreign Commerce Committee, Senator John Brick, a Republican from Ohio.The television industry, as a whole, did not trust Jones because of his previous record during his years at the Commission however, Jones immediately put together a committee to begin the investigation.   The report of the committee's findings was supposed to be finished by January, 1955.

Harry Plotnik was named assistant counsel to represent minority members of the committee in the investigation. The results of the investigation were in fact presented on February 17, 1955. The findings suggested that the networks were more in control of television than was the FCC and that the networks should be limited in their ownership of television stations.

Broadcasting magazine, in its 1971 chronology of events, lists a number of things that happened during his tenure besides that. They were: the FCC took a strong stand on any station censoring political broadcasts; it gave stations the right to editorialize;   the National Association of Broadcasters developed a new voluntary code for its members; the FCC froze TV license awards then later lifted it with a new allocation plan; the FCC repealed the AVCO rule; some networks pulled out of the NAB in a temporary dispute; TV station ownership limits were raised to seven, five VHF and two UHF; later ownership rules changed so that one person/group may own seven TV stations, seven AM and seven

FM radio stations.

In the early 1960's Jones turned to banking and became president of the People's Bank of Hanover and   president of the Pennsylvania Bankers Association. Jones died on June 23, 1963 in a hospital in Sandy Springs, Maryland, two days before his 61st birthday.

By Gerald V. Flannery, Ph.D. and Marisol Ochoa Konczal, M.S.

COY, ALBERT W.
1947-1952

Hard work, broadcast experience, and the friendship of a hometown hero were the building blocks of a successful career for Albert Wayne Coy. From humble beginnings, he grew up to influence key decisions in the Federal Communications Commission and other areas of broadcasting and government.

He was born on November 23, 1903, in Shelby County, Indiana. His parents were Lillian Monell and Albert Roscoe Coy. Coy's father, Albert, worked as a railroad station agent and supplemented the family income by taking odd jobs. The hard work of his father must have influenced Coy because at 16, while still in high school, he took a job as reporter for the Franklin Star.

After graduating from high school in 1920, he continued to report for the newspaper to pay his expenses at Franklin College. While attending the school, Coy joined Phi Delta Theta and was also elected a member of Sigma Delta Chi, the professional fraternity in Journalism. He earned his B.A. degree from Franklin College in 1926.

From 1926 to 1930, Coy worked as city editor of the Franklin Star. Then, with college degree in one hand and experience in the other, he set out on his own. In 1933, he bought the Delphi Citizen and continued as editor and publisher of the newspaper until 1933. It was during this time that Coy aligned himself with hometown hero, Paul V. McNutt. McNutt, who had been Dean of the Indiana University Law School, was running for governor in 1932. Coy worked closely with him and, when Mcnutt won the election, Coy was given a position as his personal secretary.

The position proved to be an important political stepping stone. Between the years 1933 and 1937, he held a variety of important state and regional positions. When McNutt was appointed United States High Commissioner to the Philippine Islands, Coy went with him to serve as administrative assistant from 1937 to 1939. From there, the two men next went to Washington, D.C. McNutt was heading up the Federal Security Agency and Coy was assigned the task of reorganizing the FSA.

Coy's reputation was growing in top political circles and, in 1941, he was asked to serve as a special assistant to President Franklin D. Roosevelt. During the three years he worked with the president, Coy became known as one of the most influential men in Washington.

Coy returned to his media roots in 1944 when he became assistant to Eugene Meyer, publisher of the <u>Washington Post</u>; as his assistant, Coy was made a vice-president and put in charge of the company's two broadcast properties, WINX and WINX-FM. He also served on a number of committees of the National Association of Broadcasters and, in 1946 and 1947, he headed the industry committee which cooperated with the FCC on the simplification of broadcast application forms.

On the basis of his reputation as a federal administrator and his broadcast experience, President Harry Truman appointed Coy Chairman of the FCC on December 29, 1947. Coy, who was filling the seat of Charles R. Denny after Denny resigned, was confirmed by the Senate on January 30, 1948. As a Democrat, Coy's appointment gave the FCC political balance with three Democrats, three Republicans, and one Independent.

Many key issues came up during Coy's years as an FCC Commissioner. Among them were the right of a television station to editorialize if people with opposing viewpoints were given a chance to respond, a freeze on applications for broadcast properties so the FCC could study the flood of applications waiting to be considered along with possible technical advances, the removal of the right of broadcasters to censor political speeches (but the stations would not be held responsible for libel), and restrictions on giveaway shows due to Federal Lottery Laws.

Coy served on the Commission during the early explosive growth years of television. In 1948, for example, a black and white 19" RCA console television model cost about one thousand dollars, quite an investment for the average working person, yet television gradually swept the country, coast to coast. Programs broadcast from network headquarters in New York were scheduled according to Eastern Standard Time, so an adult drama slated to air at 10 p.m. on the East coast was seen at 7 p.m. in California. It changed family social habits in many ways, one of them being the normal dinner hour at the dinner table. Families didn't want to miss the programs while they ate, so America created the "TV Tray" and meals were eaten around the television set.

After being renominated as Chairman in 1951, Coy resigned from the Commission in February 1952. He then became president and manager of KOB and KOB-TV in Albuquerque, New Mexico. The stations were owned by Time, Inc., and when the company bought WFBM and

WFBM-TV in Indianapolis, he returned to his home state to act as a consultant for the stations.

Broadcasting magazine, in its 1971 chronology of events, lists a number of things that happened during his tenure besides that. They were: the FCC took a strong stand on any station censoring political broadcasts; it gave stations the right to editorialize; the National Association of Broadcasters developed a new voluntary code for its members; the FCC froze TV license awards then later lifted it with a new allocation plan; the FCC repealed the AVCO rule; some networks pulled out of the NAB in a temporary dispute; ownership rules changed so that one person/group may own seven TV stations, seven AM and seven FM radio stations.

Coy married Grace Elizabeth Cady on September 24, 1927, and the couple had three children. Coy was a Baptist and a Mason. He died on September 24, 1957.

By Gerald V. Flannery, Ph.D. and Richard E. Robinson, M.S.

STERLING, GEORGE E.
1948-1954

Amateur radio was always George Edward Sterling's passion. Throughout his professional career he maintained an interest in and love for amateur radio. In fact, the only time he was not associated with radio was when he served in the Armed Forces. Radio became the means for Sterling to rise through the ranks to reach a seat on the Federal Communications Commission. Because he was so knowledgeable about amateur radio, his fellow Commissioners often deferred to Sterling by asking him for guidance and advice in matters affecting it.

His colleagues, at a special ceremony honoring Sterling, said:

> He brought to that office an unprecedented experience in radio, which had its beginnings in 1908, and as an operator, engineer, author and policy maker, and has played a prominent role in the development of the radio art... His warmth, sincerity and fundamental sense of fairness and affection for all of his distinguished career, long will be an inspiration and example.

One of Sterling's most widely known achievements was The Radio Manual, written in 1929 and recognized by Broadcasting magazine "as a standard textbook in radio communication, equipment and procedures by radio schools and for government training." He also gained recognition for participating in the break-up of clandestine broadcast stations established in South American countries by Axis espionage agents while he was chief of the FCC's Radio Intelligence D i v i s i o n .

George Edward Sterling was born at Peaks Island, Portland, Maine on June 21, 1894. He was son of Wesley and Annie (Tatman) Sterling. He attended public school in Maine and took special courses at Johns Hopkins University Night School and Baltimore City College. At age 18, Sterling applied for and received one of the first radio licenses issued after the passage of the Radio Act of 1912. Three years later he was serving in the Infantry and later in the United States Signal Corps in

France during the First World War. While in France he assisted in organizing and operating the first radio intelligence section of the Signal Corps. This section located enemy radio stations and intercepted their messages. After the war, he continued his association with radio by first becoming a radio operator In the Merchant marine and then becoming a marine radio inspector. He began his federal service in 1923 as a radio inspector with the Bureau of Navigation of the Department of Commerce, remaining in that position until 1935.

From 1935 to 1937, Sterling worked for the Federal Radio Commission in charge of the third radio district. In 1937, he was made assistant chief of the field division of the engineering department of the newly formed Federal Communications Commission. During the 1940's, Sterling made many contributions to the rapidly expanding radio industry. He organized and headed the Radio Intelligence Division, trained personnel in radio intelligence and techniques, and established a special branch of the FCC to handle amateur radio affairs. During this time period, he was promoted to Chief Engineer of the Intelligence Division.

Sterling was nominated to the FCC by President Harry Truman to replace Ewell Jett, who resigned to become president of the Baltimore Sun Newspapers. Sterling served on the commission for six years. To Sterling, the new developments made by the industry in the 1950's, were important and perplexing, full of unanswered questions, for both the government and the radio industry. He likened it to the early days of radio and the Hoover Radio Conferences called to deal with the mushrooming giant in 1922, 23 and 24. Because of failing health, he was forced to resign from the FCC on September 20, 1954.

Sterling was also a Commissioner who came to the FCC in 1948, an explosive growth year for television in America. In 1948, for example, a black and white 19" RCA console television model cost about one thousand dollars, quite an investment for the average working person, yet television gradually swept the country, coast to coast. Programs broadcast from network headquarters in New York were scheduled according to Eastern Standard Time, so an adult drama slated to air at 10 p.m. on the East coast was seen at 7 p.m. in California. It changed family social habits in many ways, one of them being the normal dinner hour at the dinner table. Families didn't want to miss the programs while they ate, so America created the "TV Tray" and meals were eaten around the television set.

Broadcasting magazine, in its 1971 chronology of important events, listed some things the FCC worked on in that period. They were: RCA announced the first all-electronic color tube; U.S. Senator Joseph

McCarthy demanded equal time on the networks to answer his critics and report on his expanding investigation of communists in government; the FCC proposed UHF satellite and "budget" stations; the House Commerce Committee issued a report criticizing broadcasters for running beer and wine ads; individuals and groups got the right to own seven TV stations, five VHF and two UHF; U.S. Senator Warren Magnuson (D-Wash.) accepted the Chairmanship of the Senate Commerce Committee and vowed to continue its probe of TV networks; CBS President Frank Stanton broadcast the first network TV editorial; and the FCC ended its twentieth year of operation.

 Broadcasting magazine commended Sterling for his thirty years of government service saying:

> George Sterling physically leaves the FCC... But the name Sterling is permanently inscribed in the history of communications in the United States...something he had an important part in writing.

By Gerald V. Flannery, Ph.D. and Mary Syrett, M.S.

HENNOCK, FRIEDA B.
1948-1955

The first woman appointed to the Federal Communications Commission was Frieda B. Hennock who was born in Kovel, Poland on September 27, 1904, the youngest of eight children with five sisters and two brothers. In 1910, her father brought the family to America where he entered the real estate and banking business in Brooklyn, New York. She began studying piano at the age of five, and by the time she was a teenager at Morris High School in the Bronx, she was giving music lessons and earning spending money. Her parents wanted her to be a musician but, in 1924, she announced her intention to be a lawyer, not an easy choice for a woman in those days. She worked days at a New York law firm to pay for her night classes. She graduated from the Brooklyn Law School at the age of nineteen, two years before she could legally be accepted at the bar.

Hennock began her law career with $56.00 in working capital but her actions soon gave evidence of her independent spirit and her early work showed her invention and devotion. In one case, she represented two brothers charged with first degree murder and got them both acquitted.

Later, she spent $3,000 of her own money in defending a man accused of murder in a payroll robbery case. She spent the money preparing briefs, doing a ballistics study and analyzing the architecture at the scene of the crime. However, shortly before the trial, her client escaped from prison and committed suicide. The ensuing publicity brought her more work and modest fame.

Hennock, in the late 1920's focused her efforts in the area of civil law, creating first the law firm of Silver and Hennock in 1927, then going on to win a $55,000.00 fee, a staggering sum in those days. She went independent in 1934 and in this period began her public service career. For example, from 1935 to 1939, she served as an Assistant Counsel for the New York State Mortgage Commission, working on a government study of low cost housing and participating in a lecture series on law and economics at her alma mater, the Brooklyn Law School. In 1941, she joined the law firm of Choate, Mitchell and Ely, all Republicans, not only as the first female member, but the only Democrat. Politically, she became more active, focusing her attention on getting women to participate in politics and government, championing women's rights, activities which earned her the label of "liberal."

Hennock was appointed as the first female member of the FCC by President Harry Truman and confirmed by the Senate in July 1948. Her independent attitude and concern for public service made Hennock an attractive candidate for the Commission; the Interstate Commerce committee wanted someone who was objective, had a different perspective, and was separate and apart from the radio industry. She said her only connection with radio was contributing money to President Franklin D. Roosevelt who made "Fireside Chats" on radio; however, she did say that she felt she would represent women who "made up about 90% of the listening audience."

She left her successful law career without regret, saying any lawyer had a duty to perform public service. Appearing before the Interstate Commerce Commission on May 24, 1948, prior to her appointment, she told the predominantly Republican group, "I'm against you and I always have been." In accepting the post, she gave up an annual income well in excess of her salary as a Commissioner. She often expressed her belief openly that lawyers should generally take a more active part in public affairs. Hennock's legal opinions, often written on the behalf of the minority, were viewed by other attorneys as models of legal craftsmanship. She served the full seven year term.

Hennock's presence on the Commission gave it a fairly symmetrical political balance of three Democrats, three Republicans and one Independent. In 1949, she was the only dissenting vote in the ruling to allow editorializing on radio and television saying fairness was virtually impossible to enforce. She was part of the Commission that put a freeze on issuing TV licenses and who ruled on equal time provisions. Her most important accomplishment, however, was probably her campaign for educational television: she was instrumental in getting 257 channels set aside for noncommercial use. One of her major concerns was the concentration of ownership in commercial television, a stance not popular with the industry. She also wanted political candidates to get free air time. She vigorously opposed the merger of ABC and Paramount Theaters in 1953, and, that same year, also opposed the extending of television licenses from one to three years. Even though she was outvoted, she was respected for the legal arguments she made to bolster her opinions.

President Truman nominated her for a Federal District Judgeship in 1951, which some saw as a way of getting her off the Commission; however, the New York Bar Association vigorously opposed her nomination and she was not confirmed. By the time her seven years was up, Dwight D. Eisenhower was president and he did not reappoint her. Hennock told a New York Times reporter, in 1960, that her fight against the "monopolistic forces" in television resulted in her ouster.

Hennock married William Simons, a real estate broker, after she left the FCC and returned to private practice. She died five years later in Washington, D.C., on June 20, 1960 , following surgery for a brain tumor.

The FCC adopted a resolution praising her

distinguished contribution to its work during the challenging period of television's growth. Perhaps foremost among Miss Hennock's endeavors for which she will be long remembered was her devotion and impressive championship of policies which have made possible the establishment and continuing growth of educational television... Gifted with a keen intellect and endowed with tireless energy, Miss Hennock ceaselessly and unflaggingly dedicated her extraordinary capacities to the highest goals of the nation's communications services.

Frieda Hennock's public and private careers were a monument to the American idea of freedom of choice. She came to this country from Poland with nothing and in her short lifetime established a record of accomplishments born out of her own beliefs and given life out of her determination.

A longer version of this biography will appear as an article in the book Woman's Words, Women's Voices.

By Gerald V. Flannery, Ph.D. and Peggy Voorhies, M.S. candidate.

BARTLEY, ROBERT T.
1952-1972

Robert Taylor Bartley was born on May 20,1909 in London, Texas. He was the son of Samuel Edward and Meddie Bell (Rayburn). He attended public school in Dallas and graduated from Highland Park School in 1927. He later attended Southern Methodist University from 1927-1929 in Dallas where he took courses in the School of Business Administration.

At the age of 22, Bartley moved to Washington, D.C. where he began serving as special council to then later secretary of, the research and investigative staff of the House Committee on Interstate and Foreign Commerce. Bartley was a renowned statistician who made investigations into numerous public utilities. He was involved in the investigation of the public utility holding company for the House Committee on Interstate and Foreign Commerce in Washington from 1932-1934. He was more of a researcher than a politician; more of an efficiency expert than a lawmaker.

In 1934 he supervised the preparation of reports that were instrumental in the passage of several pieces of important legislation. He played a key role in the creation of the Pipe Line Common Carrier Legislation and in the Communications Act of 1934 which gave birth to the Federal Communications Commission. This association came when he was appointed director of the former telegraph division, in charge of the regulation of telegraph land lines, cable, and radio carriers. He helped with the creation of the Public Utility Holding Company Act (SEC), the Securities and Exchange Acts and the telephone Rate Investigations Act (FCC) from 1932-1934. He was considered to be a brilliant analyst. He was the senior securities analyst for the SEC between 1937 and 1939.

His further connections with the broadcast industry began when he served as assistant to the president of the Yankee Network, Inc. in 1939 and as the war program manager for the network from 1941-1943.

After four years with that network, he joined the National Association of Broadcasters as director of war activities from 1943-1945. He became an expert in communication and its laws. He also served as secretary treasurer of the FM Broadcasters, Inc. After his five year association with the NAB, he left in 1948 to take a position as administrative assistant to the Speaker of the House of Representatives Sam Rayburn who was his uncle. He stayed in that position until 1952. Prior to the FCC, Bartley headed the team that completed the merger of FM Broadcasters, Inc. This led to his being named director of government relations during the time America was going through a transition period following World War II.

Robert Bartley was appointed to the FCC by President Harry Truman in 1952. He succeeded Paul A. Walker who served a short term due to a promotion. Walker took the seat of Wayne Coy who resigned on February 21,1952.

He entered a Commission full of problems over acquisitions of broadcast facilities. His first year on the FCC saw a lift on the freeze of new television station licensing. During the following years some of the Commissioners were accused of misconduct, undue fraternization with the broadcast industry, accepting gifts and travel expenses from those involved in the proceedings, and fraud against the government. Bartley was never tarnished by these accusations, but instead he was renominated for a second term while the investigations were going on. During his reappointment hearing the comment was made by Senator Ralph Yarborough that Bartley was a dissenter in his voting record and the FCC would not be in the mess it was if the decisions would have gone Bartley's way.

As a dissenter he voted against issues he felt strongly about. He was not an advocate of censorship but he stood firm in his belief that the Commission had the responsibility to review program performance when stations file for license renewal. Bartley voted against the merger of ABC and ITT because the merger would place a major share of the national broadcast service (especially television) under the power of an expanding conglomerate corporation...with little attention given to the local needs of the public which the broadcast operations are charged with serving.

He was always concerned about stations providing service in the public interest such as providing both sides of the issue in presenting programs. He believed in the Fairness Doctrine and tried to prevent purely "money making" in broadcasting. He assisted in preparing a report saying the FCC had the power to prevent over-commercialization and warning that the matter was under strict observance.

Robert Taylor Bartley was longest reigning member of the FCC. His tenure ran from March 6, 1952 to June 30, 1972. His appointment, first by President Harry Truman in 1952, covered twenty years and four additional presidents.   He served through two Republican Administrations and three Democratic Administrations. He served under Truman, Dwight D. Eisenhower, John F. Kennedy, Lyndon Johnson, and Richard M. Nixon.

Broadcasting magazine, in its 1971 chronology of events, lists a number of things that happened during his tenure. They were: the FCC lifted the freeze on TV license awards and introduced a new allocation plan; the FCC repealed the AVCO rule; some networks pulled out of the NAB in a temporary dispute; ownership rules changed so that one person/group could own seven TV stations and seven AM and seven FM radio stations; transistor radios were introduced; the merger of ABC and United Paramount Theaters was approved; pay-per-view television was attempted, a number of UHF stations went on the air; license renewals were extended to three years; FCC Commissioner Richard Mack was investigated for his vote on the award of TV Channel 10 in Miami; videotape was introduced into broadcast operations making reruns possible; Congress investigated payola and rigged quiz games on TV; the FCC ordered a sweeping investigation into radio and TV programming and advertising practices; the first televised presidential debates were broadcast; and FCC Commissioner Newton Minow shook up the broadcasting industry by calling television a "vast wasteland."

Bartley was the FCC Defense Commissioner from 1961-1965 where he coordinated the Commission's national defense activities with the President's office. He was Chairman of the Radio Technical Commission for Marine Services, Chairman of the U.S. delegation to the World Administrative Radio Conference for Maritime Mobil Matters at Geneva in 1967; and was awarded the Marconi Memorial Gold Medal of Achievement by the Veteran Wireless Operators Association in 1965.

Bartley married Ruth Adams of Washington in 1936. They bore three children, Robert T. Junior, Jane Bartley Odle and Thomas Rayburn.

By Gerald V. Flannery, Ph.D. and Carla P. Coffman, M.S.

MERRILL, EUGENE H.
1952-1953

Few people were better prepared to take on the job of Federal Communications Commissioner,in 1952, than Eugene H. Merrill, the son of Dr. Joseph F. Merrill. He had qualifications that set him apart. He was born in 1918, the son of a Mormon educator and took an active part in the Mormon church, trying to pattern his life in accordance with the church's doctrines. He graduated from the University of Utah in 1932 with a degree in engineering. Then, in 1935, he became the Chief Engineer for the Public Utilities Commission. This job prepared him for his later work in the Commission. His first case as Chief Engineer for the PUC involved the investigation of the rates, operations, and property values of the Mountain states Telephone and Telegraph Company.

In 1940, he became the president of the National Conference of Public Utilities Commission Engineers. He served as consulting engineer for the Utah Public Utilities Commission from 1941 to 1945.

Prior to Pearl Harbor, Merrill joined the Office of Production Management. He aided in the reorganization of its successor, the War Production Board. His assignment was communications, including radio, telephone and telegraph.

In 1943, Merrill was assigned overseas for the Foreign Economic Administration. He was sent first to Australia as a power consultant, and then to Germany as deputy. He was later promoted to Chief of Communication. His duties included allocation of radio frequencies for the occupational forces and civilian population, as well as other areas of communication.

Merrill returned from Germany in 1950 to join William H. Harrison's National Production Authority, located in Washington, and aided in its establishment.

In October of 1952 he was appointed a Commissioner of the FCC. Merrill's appointment was different from any other Commissioner ever appointed. He was a recess or "lame duck" appointee. His permanent position on the Commission depended on the outcome of the presidential election.

Merrill was appointed to the FCC on October 14, 1952, by President Harry Truman, but all pre-election polls predicted that the political party of the chief executive would change from Democratic to Republican. It was predicted that Dwight D. Eisenhower would win the election. Many Democrats as well as Republicans felt that Merrill's excellent credentials would secure him a permanent seat on the commission anyhow.

However, when the election was over and Eisenhower became the new President, Merrill's nomination was withdrawn. Legal complications now surrounded Merrill's appointment. He could have served on the Commission until 1953 legally, but with Eisenhower's selection of someone else as a successor, he refrained.

President Eisenhower Nominated John C. Doerfer of West Allis, Wisconsin to succeed Merrill. Doerfer had been chairman of the Wisconsin Public Service Commission since July, 1949. He was 49 years old at the time of his nomination.

Many say that Eisenhower's nomination was merely a political balancing of power. This was extremely important when the Commissioners had to resort to voting on certain issues. With the appointment of Doerfer, the political balance of the FCC was evenly distributed. There were three Democrats, three Republicans and one Independent.

Broadcasting magazine, in its 1971 chronology of events, lists a number of things that happened during his tenure. They were: Speaker of the House Sam Rayburn banned radio-television coverage of House Committees; CBS demonstrated its all-electronic color television system; the FCC lifted its freeze on licenses through its Sixth Report and Order and offered a new allocation plan; President Truman signed the McFarland bill, the first overhaul of the FCC since 1934, allowing the FCC to issue "crease and desist" orders like the FTC; the FCC repealed the AVCO rule; some networks pulled out of the NAB in a temporary dispute; TV station ownership limits were raised to seven, five VHF and two UHF, later ownership rules changed so that one person / group may own seven TV stations and seven AM and seven FM radio stations; the merger of ABC and United Paramount Theatres was approved; and the license award length was raised to three years.

ABC, which did not get started as a radio network until the 1940's, did not have the profitable 20 years of early radio that its competitors NBC and CBS did; thus it had a lot to do to catch up after World War II. ABC went looking for a merger that would give it the financial depth and backing it needed to get into television where it again faced the deep pockets of NBC and CBS if it wanted to truly compete. After the earlier attempt to merge with ITT was shot down and the United Paramount deal was a life saver.

Merrill left the FCC on April 15, 1953, and returned to his engineering profession. Merrill was married to Barbara McCune Musser; they had four children.

By Gerald V. Flannery, Ph.D. and Bobbie DeCuir, M.S.

DOERFER, JOHN C.
1953-1960

John C. Doerfer served as an Federal Communications Commissioner and Chairman from April 15, 1953, until March 21, 1960. His tenure spanned a period in FCC history which was noted for broadcasting controversy in the areas of international allocation of airwaves, inadequate primary radio service in certain areas, quiz show and payola scandals, and the demand for greater government control of broadcasting. And it was scandal, real or imagined, which led to his resignation as FCC Chairman in 1960. He flew to Miami and back at the expense of George Storer, who was the president of Storer Broadcasting. He also spent six days and six nights aboard Storer's yacht, where he was a guest of the broadcaster. That incident was brought to public attention as a result of an investigation by the House Special Subcommittee on Legislative Oversight. It accused him of being unduly influenced by a man whose company Doerfer was supposed to be regulating. Doerfer defended his actions by insisting that he had made arrangements to repay Storer for the airfare and that he should be able to pick his friends and those whom he associated with. "My conscience is absolutely clean," he told the House Subcommittee.. The New York Times in March 1960 quoted John Moss (D-Cal.) as referring to him a "reluctant regulator of the broadcasting industry". He clashed frequently with the House over the FCC's power to deal with rigged quiz shows, payola and other deceptive practices.

Doerfer met with President Dwight D. Eisenhower on March 9, 1960, and resigned the following day, effective March 14. Frederick Ford was elevated to Chairman following Doerfer's resignation. Doerfer came to Washington in 1953, supposedly to fill a vacancy in the Federal Power Commission. But he was nominated to the FCC by President Eisenhower and was approved unanimously by the Senate on April 15, 1953. Judge Harold M. Stephens of the District of Columbia circuit court of appeals administered the oath of office to Doerfer. He entered the office with a reputation as a firm believer in free enterprise and a man of high character and integrity, and as a Commissioner he became known for his forceful opinions on antitrust issues and the statutory authority of the FCC: he also spoke out against over-commercialization and giveaways. While serving as Commissioner, Doerfer acted as Chairman of the Committee on Pre-Hearing Procedures, was a member of the FCC's Network Study Committee and the FCC's committee to study private television intercity microwave.

Doerfer's background was in utilities. He served as Chairman of the Wisconsin Public Service Commission from July, 1949, until his FCC nomination. In Wisconsin he served as Chairman of the Public Utilities Committee in the League of Wisconsin Municipalities and helped pass effective legislation for municipal and state operations.

He chaired the Committee on Regulatory Procedure for the National Association of Railroad and Utilities Commissioners. He was also a member of the Executive Committee and the Special National Defense Committee for that organization. His civic duties included a stint as secretary of the Milwaukee Executive Club, Chancellor of the Milwaukee Barrister Club, State Chairman of the National Institute of Municipal Law Officers, a membership in the Psychiatry Board of the Milwaukee Community Chest and as a member of the Budget Committee of the Madison United Givers Fund.

Doerfer, born in Milwaukee on November 30, 1904, graduated from the University of Wisconsin in 1928 with a degree in commerce. He went to work as an accountant in the Milwaukee area upon receiving his degree and attended Marquette School of Law in his spare time. He earned his law degree in 1934, five years after he married University of Wisconsin English professor Ida M. Page.

He began a general law practice in 1934 in West Allis, Wisconsin, and continued it through the Depression until 1940, when he was elected city attorney. That same year he chaired the Milwaukee County Junior Bar Association and the West Allis Committee for Wendell Wilkie, the Republican candidate for president. He was reelected to the post of city attorney without opposition in 1944 and 1948.

In that office he specialized in cases dealing with public utilities, and that led to his appointment by Governor Oscar Rennebohm to a six-year term on the Public Service Commission. He was elected by his peers on the Commission to be Chairman just two months after his appointment.

As FCC Commissioner and Chairman, Doerfer was often at odds with Congress because he spoke out against program control and censorship as proposed by the House and Senate. Television was experiencing a tremendous boom in the 1950's, and Doerfer and the FCC had the task of overseeing the powerful new medium. His belief in the American free enterprise system led to his popularity among broadcasters, but his association with members of the industry led to his resignation from the FCC. Doerfer, who was 48, when he was named to the FCC, had two sons, John Page and Gordon Lee. He died at the age of 87 on June 5, 1992.

By Gerald V. Flannery, Ph.D. and Joe Lynch, M.S.

LEE, REX
1968-1973

President Lyndon Johnson credited Rex Lee for his transformation of American Samoa from a "Pacific Slum" into a "showplace" of the South Seas. Lee's primary tool in this metamorphosis was television. He established an educational television station in what was deemed a backward illiterate province as a means of reaching every resident. ETV was considered one of the premier facilities of its kind in the world. Certainly this accomplishment alone, during his term as Governor of Samoa, from 1961 - 1967, would qualify Lee as an excellent choice for the FCC, but his credits are not limited to that alone. Lee began his career with the government as an economist for the U.S. Department of Agriculture in Idaho from 1936 -1937. Later he served as an extension agent for the University of Idaho, 1937-1938, before returning to economic work for the Agriculture Department from 1938 to 1942 in Berkeley, California.

In 1942 he joined the War Relocation Authority for four years. He started as Executive Assistant to the Director and later became Chief of Relocation and Evacuee Property Divisions. In 1946, he moved to the Department of Territories as Assistant Director. During four years in that position, Lee was loaned to the State Department and the United Nations (1949) as a consultant on the Arab refugee problem.

Over the course of his five years with the FCC, he was known as a quiet, industrious member of the agency. When Lee did speak on his pet project, educational television, he said he liked "telling it like it was." At an address before the National Association of Educational Broadcasters in December 1973, he strongly suggested that public broadcasting's management people would not last a minute in the world of commercial television. The problems with public television, in Lee's opinion, were a lack of direction, in essence, an identity problem. He said noncommercial television needed to see itself as more than educational television but rather as public telecommunication.

Lee's criticism was not directed only at the industry. He took on the

White House and its arbitrary limit of one year grants to public broadcasting, saying that was too limiting and didn't allow future planning. The two factors, he said, which could salvage public broadcasting, were 1) long term permanent financing, and 2) no government intervention. On the other hand, he thought public broadcasting should continue to seek the kind of excellence shown in two of its programs, for television: the Hollywood Television Theater and for radio, All Things Considered. He felt outstanding programs forced commercial television and radio to upgrade their offerings, in a way the FCC could not, and it also strengthened the argument for financing for public broadcasting.

Lee was also outspoken in his concern for the concentration of media control and he often voted no in media conglomerate purchase cases. He favored new regulation to break up multimedia holdings concentrated in one market. Lee felt the FCC's "'Fairness Doctrine" was good regulation, as was the right of access rule, and that the FCC should consider antitrust laws in the formulation of communication policy.

Broadcasting magazine, in its 1971 chronology of events, listed a number of things the FCC dealt with during his tenure. They were: the final death of the International Telephone & Telegraph and ABC merger; the creation of the children's Television Workshop; the widespread and controversial coverage of racial outbreaks after Dr. Martin Luther King's assassination; the U.S.Supreme court gave the FCC jurisdiction over Cable Antenna Television (CATV); the FCC put a new freeze on AM radio applications; television was criticized for its coverage of the Democratic National Convention and the demonstrations in Chicago; the FCC ruled that stations must carry "anti-smoking spots;" the FCC was forced to look at violence on television; a new Public Broadcasting System was offered; the FCC got tough on equal employment opportunity in the industry; the U.S.Supreme Court upheld the Fairness Doctrine; the major networks underwrote a study of the joint use of their own satellite system; the U.S. House and Senate agreed on legislation outlawing cigarette commercials on radio or television; the FCC banned major networks from controlling more than three hours of programming during prime time, creating access time; the White house created the Office of Telecommunications Policy; the FCC began its investigation of commercials on children's programming; and the National Black Radio network got underway with 41 stations.

In 1973, Rex Lee resigned from the Commission to return to private life and some community service. He later served as Chairman of the Public Service Satellite Consortium, as Justice Department Solicitor General (1984), and Chairman of the Pacific Telecommunications

Council (1986).

Lee was born in April of 1910 in Rigby, Idaho to Hyrum and Elza (Fransworth) Lee. He received a B.S. from the University of Idaho (1936) and a law degree in 1964. He married Lillian Carson (1937) and had five children.

By Gerald V. Flannery, Ph.D. and Peggy Voorhies, M.S. candidate.

McCONNAUGHEY, GEORGE C.
1954-1957

George Carlton McConnaughey was born in Hillsboro, Ohio, on June 9, 1896. McConnaughey, of Scotch-Irish descent, obtained his primary education from local public schools in Hillsboro. From 1914 until 1917 he attended Denison University when his college education was interrupted at the start of World War I. During the war he enlisted in the Army and became a first lieutenant of field artillery. He was in combat service in the Meuse-Argonne offensive and served as liaison with the infantry in the Battle of Verdum. In 1919, after being discharged from the Army, McConnaughey re-entered Denison University and obtained a Ph.D. degree in 1920.   To further his education, he decided to join the Western Reserve University Law School from which he graduated with an LL.B. in 1923. In 1924, only one year after his graduation from law school, McConnaughey was admitted to the Cleveland Bar. This was also the year in which he married Nelle Louise Morse with whom he had two children, George C. Mcconnaughey, Jr., and David C. McConnaughey.

From 1926 to 1928 he served as Assistant Director of Law for the City of Cleveland. During this period he also served as President of the Ohio-Michigan Province of Phi Delta Phi legal fraternity. From 1928 until 1930 he served as President of its National Council.

On February 1, 1939, he was appointed Chairman of the Ohio Public Utilities Commission and served in that post until 1945. Parallel to this position, McConnaughey also held the Chairmanship of the War Transportation Commission of Ohio (1942-1945). He served as the commission's representative on the Interstate Cooperative Council of State Governments. In 1944, he was elected President of the National Association of Railroad and Utilities Commission for a one-year term.

In 1945, he resumed his law practice in Columbus, Ohio, and on November 30, 1953, President Dwight D. Eisenhower appointed him Chairman of the Defense Contracts Renegotiation Board, a body that handled settlement of government contracts. He held that office until the following year, when the President   called on him again for public service. On September 26, 1954, he was sworn in as Chairman of the Federal Communications Commission. He succeeded Rosel H. Hyde, whose designation as chairman had expired on June 30 of that same year. The fifty-eight-year-old McConnaughey also entered the FCC in the capacity of Commissioner succeeding George C. Sterling. Sterling, whose term as Commissioner expired on June 30, 1957, retired from the FCC on September 24, 1954 because of ill health.

McConnaughey's dual appointment brought the FCC to its full strength of seven members, four Republicans (including McConnaughey) and three Democrats. His seven-year term was supposed to expire on July 1, 1956 but he served a total term of 33 months that ended in July 30, 1957.

On March 10, 1955 McConnaughey was appointed by a party line vote to the Senate Interstate Commerce Committee. The approval of his appointment caused a serious controversy led by Lyndon Johnson. At the time, his designation was considered strictly a political move by President Eisenhower. However, despite all the political unrest, his appointment was confirmed by the Senate five days later, on March 15, 1955.

Shortly after taking office, a government wiretapping case arose and he stated that the FCC would join in any effort to push for legislation against the use of wiretapping.

On February 27, 1956, five months before his term in office expired, President Eisenhower redesignated him as FCC chairman. Later that year, on September 26, a major step forward in the telephone industry took place; the first trans-Atlantic telephone line was open for use.

On June 4, 1957, McConnaughey declined reappointment for a new term on June 30. On June 26, the White House announced his resignation and a new Chairman was appointed. The Chairmanship was passed on to John C. Doerfer.

The 33-month tenure of George McConnaughey as Chairman of the FCC was a quiet time for that agency, but soon after he resigned, scandals erupted over the granting of licenses for television stations in Miami, Pittsburgh and Boston. Congress charged the Commissioners with conducting "a too informal" relationship with the radio and television Industry. In testimony before the House subcommittee on Legislative Oversight, and later in hearings conducted by the FCC, it was stated that McConnaughey and other Commissioners often had lunch and other private contacts with applicants for licenses. The most sensational charge brought against the former Chairman himself was that he solicited bribes of $50,000 from one Pittsburgh applicant, and of $20,000 from another. However, these charges were never substantiated and McConnaughey emphatically denied them under oath. As a result of this scandal, the Justice Department established a rule intended as a guideline for the FCC, namely any applicant who approaches an official of a regulatory agency secretly, outside its formal proceedings, should automatically lose the case.

Even though McConnaughey's time as Chairman was a quiet one for the FCC, some developments in the communications industry during that period are worth mentioning. FM broadcasters were authorized to engage in supplemental "functional music" operations. A study of network operations was initiated. Rule-making proceedings to consider the problems of Ultra High Frequency were initiated. The Commission called a public conference to consider the technical problems of UHF and as a result an industry committee known as TASO was developed. This organization made allocation studies for more than two years and reported important data to the Commission in 1959.

Broadcasting magazine, in its 1971 chronology of events, listed a number of other things that happened during his tenure. They were: the FCC boosted the number of television stations one entity could own to seven, five VHF and two UHF; U.S. Senator Warren Magnuson (D-Wash.) accepted the chairmanship of the Senate Commerce committee and continued its probe of the TV networks; the House Un-American Activities Committee announced a hearing on communists in radio-television; videotape recorders got more widespread use in the industry; and a film of a Denver murder trial renewed interest in radio and television coverage in the courtroom.

After McConnaughey's term expired at the FCC he started practicing law in Columbus, Ohio with his son George, Junior.

He died of cancer on March 16, 1966. He was 69 years old.

By Gerald V. Flannery, Ph.D. and Marisol Ochoa Konczal, M.S.

MACK, RICHARD A.
1955-1958

Richard A. Mack was born on October 2, 1909 in Miami, Florida and he attended Dade County public schools. After high school, Mack attended the Northwestern Military and Naval Academy at Lake Geneva, Wisconsin; then he enrolled at the University of Florida, graduating with B.S. and B.A. degrees in 1932.

After college, Mack began his career as a general insurance agent in Tampa, then moved to credit manager with the General Motors Acceptance Corporation in Miami from 1935-1940. The next two years of his life were spent as credit manager or the Hector Supply Company also located in Miami.

In 1942 he entered the Army infantry as a Second Lieutenant by reason of his R.O.T.C. commission. He served in the European theater and rose to the rank of Lieutenant Colonel. Released from active duty in 1946, he declined a commission in the regular Army but continued in the Officer Reserve Corps.

Back in civilian life, Mack became general manager for Port Everglades, Florida and was appointed by Governor Caldwell to the Florida Commission in September 1947 to fill an unexpired term. In 1948 and 1952 he was again appointed to the Florida Commission. He later became Chairman under the rotation system.

Mack served as a member of the Executive Committee of the National Association of Railroad and Utilities Commission and was elected its Second Vice President in 1954; and he assisted ICC Commissioner James K. Knudson when the latter became Administrator of the Defense Transportation Administration.

His position with the NARUC brought him into close association with the Federal Communications Commission. He represented the NARUC on ICC and FCC panels. On the Florida commission, he twice participated in FCC common carrier hearings--in 1949 and 1950, on the Bell System's acquisition of Western Union's long distance facilities, and in 1951, on a Western Union rate increase. President Dwight D. Eisenhower nominated Mack to the FCC and he took his oath on July 7, 1955.

Mack's term on the FCC was a stormy one. During his tenure, the Special House Subcommittee on Legislative Oversight began investigating regulatory agencies and the FCC. Its chief council, Dr. Bernard Schwartz, told the subcommittee the next thing he was going to investigate was the February 1957 award of TV Channel 10 in Miami to Public Service TV Inc., a National Airlines subsidiary. In granting the channel to the airlines subsidiary, the FCC had reversed the findings of its own hearing examiner, Herbert Sharfman, who had recommended it go to WKAT, Inc. of Miami Beach. The subcommittee had announced that it would question FCC member Richard A. Mack, a democrat, appointed by President Dwight D. Eisenhower. After a secret session on February 10, in which Schwartz accused most members of seeking to "whitewash" his investigations, he was dismissed by a 7-4 vote.

Schwartz told the Special House Subcommittee on Legislative Oversight that Mack, 48, had received $3,650 from Thurman A. Whiteside, alleged "fixer" for a National Airlines subsidiary. Schwartz, the committee's ousted counsel, was testifying in answer to a challenge that he substantiate his charges against FCC members. Four votes of the seven-member FCC were needed to award a channel. Originally, only three Republican members favored Public Service.That was not enough to award the license. The fourth vote came from Mack.

Mack voted for Public Service despite the fact that he acknowledged to two subcommittee staff attorneys that he personally knew several key officers of the firm had rather "shady backgrounds." Mack described money he had received from Thurman Whiteside as "loans" and admitted that Whiteside had "forgiven" a "portion " of them.. Whiteside said he had been Mack's friend for 40 years, had lent him money "since we were 18 years old" and always had been repaid. The balance of the loans "stood at $250," Whiteside asserted. He denied that Mack was pledged to him.

Mack asked the subcommittee for an early chance to "answer the un-substantiated charges... so recklessly made" by Schwartz. He said he was confident the subcommittee would conclude that they were "without foundation."

Thurman Andrew Whiteside, 48, Miami lawyer and Mack's boyhood friend, admitted to the House subcommittee February 24-26 that he had (a) lent Mack $7,850 in the past seven years, (b) given Mack a 1/6 interest in the Stembler-Shelden Insurance Agency, Inc. of Miami, from which Mack received $9,822 in profits in 1953-56 and (c) made Mack sole stockholder (at no cost to Mack) of Andar, Inc., a holding company "licensed in the insurance business," from which Mack received $4,350, of which $2,000 was used to repay some of Whiteside's loans to Mack.

Whiteside conceded February 24 that he had visited and phoned Mack repeatedly for eight months to urge him to give the National Airlines application for Channel 10 "every consideration," but he denied that he actually had represented the company. He said he had rejected a $10,000 fee offered by his ex-law partner, Florida Circuit Court Judge Robert H. Anderson, to take the case.

Richard A. Mack,48, resigned his $20,000-a-year FCC post on March 3 in the midst of a Congressional inquiry into his activities. Mack had insisted in testimony before the Special House Subcommittee on Legislative Oversight February 27-28 that he not been influenced improperly as an FCC member. But he confirmed having received loans from an old friend who was interested in the FCC award of a valuable Miami TV channel. Most members of the subcommittee called for Mack's resignation.

Mack Told President Eisenhower in his resignation letter that he was quitting " voluntarily" because " my usefulness" as an FCC member "has been brought into question." He asserted that " I have not violated my office in this instance (the Miami case) or in any other during my career in public life." Eisenhower, immediately accepted Mack's resignation, with a letter saying: "Without attempting to pass judgement...I nevertheless agree with you that your usefulness as a member of the commission is...seriously impaired."

Ex-FCC member Richard A. Mack and Miami lawyer Thurman A. Whiteside were indicted by a federal grand jury in Washington September 25 on charges of conspiring to defraud the United States In Miami's TV Channel 10 case.The 14-week trial ended in mistrial July 10, 1959 with the jury deadlocked at 11-1 for conviction. A federal jury in Washington acquitted Whiteside of conspiring with Mack. Whiteside also was cleared of trying to influence Mack. Mack, under treatment for Alcoholism, was declared unfit to stand trial.

Mack, 54, was found dead in a Miami rooming house November 26, 1962.

By Gerald V. Flannery, Ph.D.

FORD, FREDERICK W.
1957-1964

Frederick Wayne Ford came to power in the wake of scandal in the broadcasting industry. Ford stepped into the spotlight in 1960 with promises that, in the future, the Federal Communications Commission would play a more vigorous role in the business of broadcasting.

Ford was born on September 17, 1909, in Bluefield, West Virginia. He was the only son of George M. and Annie Laurie Ford, both of whom worked in the public school system. He attended public schools in Charleston and Dunbar, West Virginia, and went on to West Virginia University. He graduated from there in 1931 with a B.A. degree and received his LL.B. degree from that university's law school in 1934. While a law student, he was a member of the editorial staff of the West Virginia University Law Quarterly.

Ford began the general practice of law before state and federal courts in 1934 as a junior partner in the firm of Stathers and Cantrall of Clarksburg, West Virginia. He was there for five years before entering government service. In 1939, he began work in the Office of the General Counsel of the Federal Security Agency which, at that time, was responsible for the administration of Social Security benefits.

He transferred to the legal staff of the Office of Price Administration in 1942, but, in that same year, entered the armed forces as a Second Lieutenant in the Air Force. He served until 1942, at which time he left the service with the rank of Major. Returning to the Office of Price Administration, he became a Hearing Commissioner.

In 1947, he transferred to the Federal Communications Commission, where he worked in the hearing and review sections. He served the FCC in the special technical and legal group and in the general counsel's office. In 1950, he was named an FCC trial attorney and became the first Chief of the Hearing Division of the Broadcast Bureau when the Commission was reorganized in 1951.

Leaving the FCC to join the Department of Justice in 1953, he began as a First Assistant in the Office of the Legal Counsel and later as

Acting Assistant Attorney General. In January 1957, he became Assistant Deputy Attorney General under Deputy Attorney General William P. Rodgers, who later became Attorney General of the United States.

On August 29, 1957, President Dwight D. Eisenhower appointed Ford to fill the seat of retiring FCC Chairman George C. McConnaughey. While a member of the FCC, he acted as liaison with the Office of Civil and Defense Mobilization in long-range frequency allocation planning. He also served as an alternate Commission member of the Interagency Telecommunications Advisory Board, which advised the Director of Defense Mobilization in matters relating to national telecommunications plans. Additionally, he served as an FCC Alternate Defense Commissioner and as a member of the Commission's Telephone and Telegraph Committees.

Ford's predecessor, John C. Doerfer, had run the agency with a "hands-off" attitude toward regulation and resigned under fire for accepting "unusual hospitality" from a broadcaster. Ford was determined to meet what he saw as FCC problems by adopting a firm policy toward television and radio licensees. Major issues during his Chairmanship include the use a of temporary license suspension, stricter definition of permissible contacts between Commissioners and broadcasters, and the licensing of networks.

Ford was on the FCC when President John F. Kennedy was assassinated and the television networks launched the most massive coverage of everything connected with that event. Work stopped all across America and citizens listened or viewed the early coverage in shock and sadness. Gradually, they filtered back to their jobs but continued to follow the coverage, often with tears in their eyes, crying openly as they went about their business. Scholars say that television coverage allowed people to become part of the family, to witness the shock, the sadness, the sorrow, to live through the funeral experience, just as if it was someone close to them. The impact of television was clearly seen and felt.

Broadcasting magazine, in its 1971 chronology of important events, listed a number of things the FCC dealt with in that period. Some of them were: the FCC began a three year experiment with on-air pay television; it issued its Barrow Report calling for 37 changes in TV network practices; it approved a "patron" plan whereby businesses could underwrite a ETV program without violating the no sponsorship rule; the use of videotape spread; the payola and quiz show scandals swept the country; a satellite sent weather reports to earth from 400 miles in space; the FCC imposed a partial ban on AM licenses; the FCC warned broadcasters they might have to prove their ratings claims.

At the end of his term with the FCC, Ford himself was involved in a small controversy. He had turned in his resignation effective December 31, 1964; however, the White House failed to report Ford's departure until mid-January. The delay was embarrassing to Ford because he had taken a position as President of the National Community Television Association. The lack of a formal announcement by President Lyndon Johnson meant that Ford represented a trade association with vital interests in Commission proceedings while still a Commissioner.

Ford left the FCC having accomplished major changes. He calmed criticism of the industry, developed stronger policy toward station licensing, and encouraged broadcasters to consider the community when programming.

By Gerald V. Flannery, Ph.D. and Richard E. Robinson, M.S.

CROSS, JOHN S.
1958 - 1962

He came. He saw. He battled. These words along with the quote found in Broadcasting magazine (Nov. 15, 1976), " I may not be awfully smart, but I'm as clean as a hound's tooth. . ." form a fairly accurate description of John Storrs Cross, Federal Communications Commissioner and a near 45 year veteran of the Washington officialdom.

Cross was swept from the black-and-white world of the Department of State's Telecommunication Division into the colorful war zone, which was the FCC, on May 15, 1958. Representative Oren Harris, Chairman of the House Subcommittee on Legislative Oversight, was spear-heading the investigation of, the soon-to-be, ex-Commissioner Richard A. Mack and his ex parte contacts. That investigation finally led to Mack's resignation on March 3, 1958, which opened the door for Cross.

Commissioner Mack's conflict of interest was characteristic of the late 1950's. Scandals, quiz show fixing, subliminal advertising and payola schemes plagued broadcasters and caused large amounts of consternation within the halls of Congress. Into all of this came John Storrs Cross to hopefully calm things down and at least alleviate the fears of corruption in the FCC.

Even within the FCC, the age old bureaucratic excuse was heard, "Well, it looked good on paper." Not everybody was happy with Cross' appointment, one of the most vocal having been Dr. Bernard Schwartz. Schwartz, who was a professor on constitutional law at New York University, had just recently been ousted as Chief Legal Counsel to the House Subcommittee on Legislative Oversight because of his outspoken criticism on the Subcommittee's previous inaction on the Mack case. Echoing the feelings of others who felt that Cross, who called Arkansas home, would be nothing but Arkansas Representative Harris' "boy" on the FCC, Schwartz said in The New York Times article, "The President has missed an 'excellent opportunity' to exercise his prime responsibility for appointing men of high caliber to vital agency positions."

While it seems that neither side was correct, the Schwartz faction appeared to be a bit closer to the truth. According to James L. Baughman in his book, Television's Guardians: The FCC and the Politics of Programming, 1958-1967, "Cross often voted against [FCC Chairman Newton] Minow and rarely kept up with the workload."

The irony behind that statement was that one of the few times that Cross supported Chairman Minow was for a plan to reorganize the FCC. This plan, in theory, would have lightened the individual Commissioner's workload. However, a majority of Congress felt that the plan gave the FCC Chairman too much power, and therefore voted 323-to-77 against its adoption.

Cross generally opposed Chairman Minow's proposals; one of the most exemplary being the Channel 13 decision; being the only dissenting vote in a 6-to-1 decision to allow Educational Television for the Metropolitan Area, Inc. to buy the New Jersey station, WNTA-TV.

Cross defended his action by stating to The New York Times that

he was a 'firm believer' in educational television. But . . . even this interest was not sufficient to ride roughshod over the e x p r e s s e d opposition of the duly elected officials of New Jersey w i t h o u t giving them an opportunity to be heard. He went on to   p o i n t out that according to the Communications Act the Commission in distributing licenses, must make a 'fair, efficient and equitable distribution' of service among the several states and communities.

This defense is ironically interesting coming from the man who stated in The New York Times , prior to his appointment, that "they've [FCC] got enough lawyers over there already . . . I think it'll be an asset to me not being one. And anyhow, I'm a reasonably good sea lawyer. I can sling the lingo around if I have to."

John Cross was born in Birmingham, Alabama on September 18, 1904, to Thomas and Elise (Troy) Cross. He attended the McCallie School for Boys in Chattanooga, Tennessee and the Marion Military Institute in Marion, Alabama. He received a Bachelor of Science degree in Electrical Engineering from Alabama Polytechnic Institute in 1932. Despite all of this, Cross always claimed Arkansas as his home.

Early in his career, he worked for Studebaker Corporation in their engineering laboratory. He also worked for the Realty Trust Company of Detroit and as a construction superintendent for the S.S. Kresge Company.

At the same time, Cross caught the investment fever of the late 1920s. He was going to realize his American Dream, retirement at age 30 as a millionaire. In 1929, he was well on his way with two cars,

$100,000, and an Irish hunting horse; then the market came tumbling down taking his dream with it.

Cross, for a time, seemed to bounce from one job to another across America. He worked for the Michigan Highway Department for a short while, before trying his hand, unsuccessfully, at being a newspaper columnist in South Carolina. In 1931, he began work for the Department of the Interior as a field engineer for the National Park Service.

This began his career in electrical communications. From 1942 to 1946, he helped the U. S. Navy increase its world-wide wartime communications four-fold as a Naval officer. He helped build radio stations and train the personnel to operate them.

Afterwards, he went to work as the Assistant Chief of the State Department's Telecommunications Division. While at the Department of State, Cross attended approximately eighteen international conferences on such subjects as frequency allocations and radio calls.

Cross then worked as a Commissioner for the FCC 1958 to 1962. He was considered for reappointment to the FCC and could count on the support of Representative Oren Harris, both Senators from Arkansas, John McClellan and J. William Fulbright, but he did not get reappointed. President John F. Kennedy instead appointed E. William Henry, a long time friend of his brother Robert. Robert Kennedy had been hoping to name Henry to almost any independent agency since 1961. From that point until he died, Cross worked as a communications consultant specializing in dealings with the FCC.

John Storrs Cross died on November 8, 1976, of a heart attack, heading from his summer home in Eureka Springs, Arkansas, to Bethesda, Maryland. He was survived by his wife Mary (Fuller) Cross, daughter of former Representative Claude Fuller of Arkansas, and his two sons, John Fuller Cross and Capt. Claude Christopher Cross, USN.

John Storrs Cross. He came. He listened. He served well.

By Ken Ditto, M.S. and Gerald V. Flannery, Ph.D.

KING, CHARLES H.
1960-1961

The late 1950's were trying times for the Federal Communications Commission as the FCC Chairman, John C. Doerfer, was forced to resign under pressure from President Dwight D. Eisenhower. A Congressional investigation indicated that Doerfer was too friendly with George Storer, the head of Storer Broadcasting Company which owned five television and seven radio stations.

Doerfer was not the first major FCC member to step down during the Eisenhower presidency, therefore potential replacements came under great scrutiny. A New Jersey lawyer, Edward K. Mills, was nominated, but declined because of a possible conflict of interest. The seat was filled in July of 1960 by Charles H. King in an unconfirmed appointment. King served on the FCC from July 1960 to March of 1961 under Eisenhower's approval only because the Senate was in recess. His Republican political presence established a majority on the restored seven- member Commission.

King's short term on the FCC resulted in his participation in only a few important votes. For example, in September 1960, his voted yes on a ruling concerning network control of prime time and the FCC created a new regulation about it. Affiliate stations were now able to reduce network option time and reject network programs in favor of local or national programs more suitable to their audience. A new regulation requiring that public notice be given when broadcast facilities were for sale also came during his term.

In January 1961, the newly elected, President John F. Kennedy, replaced King with Newton Minow, a Democrat. King remained with the Commission an additional three months during the transition period. Minow went on to chair the FCC, ousting then head, Frederick Ford.

King's background indicates that Eisenhower selected him for the post on the FCC because he was not affiliated with any radio or television businesses, and therefore was not beholden to, or friends with, anyone who might influence him. When asked to serve, King was the Dean of the Detroit College of Law, a position he had held since 1944. During his term on the Commission, he took a one year leave of absence from the school, a clear indication that his political ambitions were limited.

A well respected lawyer, King continued his law practice during his teaching years. He was known for his work with appellate cases before his native Michigan Supreme Court. King also authored numerous articles for law periodicals which included a review of Michigan law during war time and he provided information to update lawyers returning from the war on state legal matters.

King joined the Michigan Bar in 1933 after graduating from the Detroit College of Law the same year, financing part of his expenses by writing a book An Outline to Evidence which was used in his studies. In 1940 he received an A.B. from the Detroit Institute of Technology and a Masters in law from the University of Michigan in 1941. King's teaching career began in 1933 when he taught part-time at the Detroit College of Law while practicing as an attorney. In 1937 he began a full-time teaching career and by 1944 became Dean of the school.

In the 1950's, King entered public service. From 1951 to 1953, he represented the Highland Park, Michigan community as Civil Service Commissioner. In 1952, he chaired the Michigan "Taft-for-President" committee; then ran unsuccessfully for justice of the Michigan Supreme Court. King also served as President of the Michigan Law Institute, was a member of the Detroit, Michigan and American Bar Associations, the American Law Institute and the American Judicature Society.

Charles Henry King was born in Gulfport, Mississippi on August 8, 1906, to William A. and Gertrude (Cody) King. The family moved to Bradford, Pennsylvania where he received his elementary education, then to Highland Park, Michigan for his secondary schooling. He married Ouida A. Jolls in 1933 after law school. They had one son, Charles H. King, Junior.

A respected educator and attorney with much integrity and little political ambition, King was the perfect candidate for President Eisenhower's troubled FCC. Although he served only a brief nine months, King accomplished what he was chosen to do -- vote objectively and keep things running smoothly until both he and the President left their respective posts.

Broadcasting magazine, in its 1971 chronology of events, listed things the FCC dealt with during his tenure. They were: the use of a satellite to send weather reports back to earth from 400 miles in space; the Federal Trade Commission investigated broadcast ratings and how they were used in advertising; the FCC established the Office of Complaints and Compliance to keep an eye on broadcasters; the FCC started keeping a close watch on broadcasters to see how they met community needs; the Kennedy / Nixon debates drew a huge audience and visibly affected the campaign; the FCC moved to stop station trafficking by requiring a license be held a minimum of three years before selling it; and the FCC approved FM stereophonic broadcasting.

By Gerald V. Flannery, Ph.D. and Peggy Voorhies, M.S. candidate.

MINOW, NEWTON
1961-1963

Like many of President John F. Kennedy's appointees to federal office, Newton N.Minow had more youthful energy and natural intelligence to support him than he did time honored experience when he became Chairman of the Federal Communications Commission. Sworn in on March 2, 1961, Minow, a Democrat, was then 35 years old.

It would not be long, however, before the young attorney would make what perhaps today remains the most famous speech ever given by an FCC Commissioner. In May 1961, just two months after taking up the Commission reins, the scholarly, bespectacled Minnow grabbed national front-page headlines when he told a shocked gathering of broadcasters that the television programming that they had spawned had created a "vast wasteland." Although the speech ultimately did little to inspire improvements in television programming, it survives today as a testament -- stunning in its stark frankness and stirring in its simple eloquence -- to Minow's belief that television programming must provide more than merely bubblegum for the eyes.

"When television is good," Minow wrote in his speech,

nothing, not the theater, not the magazines or newspapers, nothing is better. But when television is bad, nothing is worse. I invite you to sit down in front of your television set when your station goes on the air and stay there without a book, magazine, newspaper, profit and loss sheet, or rating book to distract you, and keep your eyes glued to that set until the station signs off. I can assure you that you will observe a vast wasteland. You will see a procession of game shows, violence, audience participation shows, formula comedies about totally unbelievable f a m i l i e s, blood and thunder, mayhem, violence, sadism, murder, Western badmen,Western good men, private eyes, gangsters, more violence and cartoons. And,endlessly, commercials -- many screaming, cajoling and offending. And, most of all, boredom. True, you will see a few things you will enjoy. But they will be very few. And if you think I exaggerate, try it. Is there one person in this room who claims that broadcasting can't do better?

Minow went on in his speech to outline his philosophy as Chairman of the FCC, and he implored television broadcasters to consider the needs of the public when making programming decisions.

I urge you to put the people's airwaves to the service of the people and the cause of freedom," Minnow concluded. "You must help prepare a generation for great decisions. You must help a great nation fulfill its future. Do this, and I pledge you our help.

Minow pronouncements began a new era in broadcasting regulation -- involvement in programming.

Minow was born on Jan. 17, 1926, in Milwaukee, Wisconsin, the son of Jay and Doris (Stein) Minow. He was reared in the Jewish faith and attended public schools, enrolling after high school at the University of Michigan. His college education was interrupted by World War II, and he served in the U.S. Army Signal Corps until 1946, achieving the rank of Sergeant. After military service, he returned to college, this time enrolling at Chicago's Northwestern University, from which he received a Bachelor of Arts degree in Speech, graduating Phi Beta Kappa in 1949. A year later, he earned a law degree from Northwestern. While a law student, he was chosen editor-in-chief of the school's law review. He graduated first in his class, was a member of the legal fraternity, Order of the Coif, and was the recipient of the Wigmore Award as the senior who had contributed most to the law school's reputation.

Minow was admitted to the bar in Illinois and Wisconsin in 1950 and began his law career in Chicago. In 1951 he was appointed law clerk to Chief Justice Fred M. Vinson of the United State Supreme Court. A year later he became administrative assistant to then Governor of Illinois Adlai Stevenson. Minow was a active campaigner for Stevenson in his bids for president in 1952 and 1956. It was during the second campaign that Minow travelled with Robert F. Kennedy, a relationship that eventually, In the 1960 presidential election, led Minow to become the secretary and general counsel of the National Business and Professional Men and Women for Kennedy-Johnson, and chairman of Citizens for Kennedy in the suburban Chicago.

It was on January 9, 1961, just eight days before Minow's 35th birthday, that President Kennedy announced his intention to appoint

Minow to the FCC. Kennedy made the nomination on January 30, and Minow was confirmed by the U.S. Senate on February 30 for a term to run until June 30, 1968. His appointment made the balance on the FCC 4-3 in favor of the Democrats as he took the seat of Republican Charles H. King and replaced Republican Frederick W. Ford as Chairman.

Minow's tenure was to be flavored with "vigorous application to the law," according to the April 10, 1961, issue of <u>Broadcasting</u> magazine, particularly in light of revelations of "rigged" quiz shows and "payola scandals" that prompted increased FCC moves to punish violators of Commission regulations. Lawrence W. Lichty wrote of Minow in <u>American Broadcasting</u> (1975):

> His early concern with the legality of the FCC's programming requirements in light of his legal background and his expressed concern over educational broadcasting (following his association with an educational film production company and Midwest Council for Airborne Television) seem to bear out the general thesis of this [stricter enforcement]. It is clear that Minow appointment tipped the balance in favor of tougher regulation.

Whatever Minow's appointment meant for an FCC under his stewardship was to come to an end a bit shorter than expected. He did not complete his seven-year term and left June 1, 1963. Minow returned to the private sector, taking up the post of executive vice president, general counsel, and director of Encyclopedia Britannica, Inc. until 1965. In that year, he became a partner of the Chicago law firm of Sidley & Austin. He had a general practice, but also specialized in communications law.

After leaving the FCC, Minow held many and varied positions on the boards of both private and public organizations, including Public Broadcasting Service, Rand Corporation, Pan Am World Airways, and the Mayo Foundation. He also was an author, having written <u>Equal Time: The Private Broadcaster</u> and the <u>Public Interest</u> (1964), co-authored <u>Presidential Television</u> (1973) and contributed to <u>As We knew Adlai</u> (1966). He continued to contribute to legal journals and magazines. Minow received a variety of awards, among them the George Foster Peabody award (1962), the Dr. Lee DeForest award of the National Association for Better Radio and TV (1962), and the Ralph Lowell award (1982).

Minow lived in Glenco, Illinois, with his wife, the former Josephine Baskin of Chicago.

By Michael A. Konczal, M.S. and Gerald V. Flannery, Ph.D.

HENRY, EMIL W.
1962-1966

Emil William Henry was born in Memphis, Tennessee on March 4, 1929 to Elizabeth T. and John Phillips Henry. He attended a local elementary school and later went on to the Hill Preparatory School in Pottstown, Pennsylvania. He graduated *cum laude* from Hill Preparatory, then enrolled at Yale University. He received a B.A. in 1951 and, after his duty in the Korean campaign, he received an LL.B. degree in 1957, from Vanderbilt School of Law in Nashville, Tennessee.

While at Yale, Henry got some radio experience as a member of the student organization which operated WYBC, the "indoor" broadcasting system wired to each dormitory. He worked in the continuity department and also was an announcer. At Vanderbilt, Henry was associate editor of the university's Law Review and was elected to the Order of the Coif, the national honor society of the legal profession.

After completing Officer Candidate School in 1951, he was commissioned an ensign in the Navy. He saw active duty for three years aboard the U.S.S. Bausell, a destroyer with the Pacific fleet, as a lieutenant; he was discharged in 1954. Prior to the completion of his duty, he had served as the U.S.S. Bausell gunnery officer.

After receiving his law degree in 1957, Henry was admitted to the Tennessee State Bar that same year and practiced in both state and federal courts. He later became a partner in the firm of Chandler, Manerie and Chandler. He was elected as a member of the bar of the United States Supreme Court in 1960, and during that year, while attending a bar association function, he became acquainted with Robert F. Kennedy. He accepted an invitation to work in the Kennedy camp, and was appointed representative to the minorities division of the Democratic National Committee. Henry took charge of the campaign among minority groups. After John Kennedy's election, Henry returned to Memphis to practice law. In late August, 1962, President Kennedy nominated him for a seven-year term to the Federal Communications Commission.

Henry was sworn in on October 2, 1962, and according to Lawrence Laurent, television critic with the <u>Washington Post</u>, it was the quickest, smoothest, and least argumentative confirmation hearing in recent FCC history. When Newton Minow revealed that he planned to resign in May 1963, Kennedy announced his intention to name Henry as the Commission's next Chairman.

On June 2,1963, Henry officially took over the Chairmanship of the FCC. He was 34 at the time, one of the youngest men ever to head the Commission. He acknowledged seeing some "green shoots" in Minow's famous speech about the "vast wasteland," that was television, and he planned to consolidate some of the gains Minow had secured. He appeared on June 20, 1963, before a House special investigations sub-committee looking into radio and television ratings problems.

Henry was instrumental in the Commission proposal to reduce the influence of the three major networks on evening TV programs. A rule was inaugurated defining "option time" (air hours) which the networks could commandeer from their affiliated stations for network programs. Henry was not a controversial Chairman, usually he pointed out broadcasting's shortcomings with wit and tolerance and was generally tolerated by broadcast officials in return.

Broadcasting magazine, in its 1971 chronology of important events, lists several things the FCC dealt with during his tenure. The were: making FM allocations; the Federal Communications Bar Association decided to try to rewrite the 1934 Communications Act; the FCC tried to find a way to limit amount of time devoted to commercials on television; a government investigation said TV violence and juvenile delinquency were definitely related; and the FCC said it had authority over all CATV.

Before Washington, Henry was prominent in civil rights and served on the Tennessee advisory committee to the U.S Commission on Civil Rights. In Memphis, he was director of the local chapter of the American Red Cross, vice-president of family services, general counsel of the Tennessee-Arkansas-Mississippi Girl Scouts, and he belonged to the Tennessee Bar Association and the American Bar Association.

On April 8, 1966, Emil William Henry resigned as Chairman of the Federal Communications Commission. After Leaving the FCC, Henry campaigned in the Democratic gubernatorial primaries for John J. Hooker, Junior. Henry also joined Arnold and Porter, attorneys at law, in Washington, D.C.

In 1967, Henry became the Director of the National Association of Educational Broadcasters. It was during this time that he was also a member of the National Citizens for Public Television. The committee's goal, at that time, was to stimulate support for a system of non-commercial television with diversified program service.

Henry campaigned to re-elect New York Mayor John Lindsay in

1969. In 1970, he joined some former high level officials in a fight against the Internal Revenue's efforts to deny tax exempt status to organizations that file lawsuits in the public interest.

Henry married Sherrye E. Patton, and they had three children.

By Gerald V. Flannery, Ph.D. and Bobbie DeCuir, M.S.

Kenneth A. Cox was born in December 1916 in Topeka, Kansas; after moving several times, Cox graduated from High School in 1934 in Seattle, Washington. He attended the University of Washington and received a bachelor's degree in 1938 and a law degree in 1940. The following year, he received his master's degree in law from the University of Michigan. Cox, in 1941, began working as a law clerk for Justice William J. Steinert of the Washington Supreme Court, and later worked for the State Attorney General as a lawyer on the staff of the State Tax Commission.

When World War II broke out, Cox's legal work was interrupted and he entered the Army, in 1943, as a private with the Quartermaster Officer Training School. He was later commissioned and became an instructor. In that same year, he married Nona Fumerton of Seattle and they had three sons. Cox was assigned to the Pentagon where he edited intelligence reports; in 1946 he left the Army with the rank of Captain.

After the war, he returned to the University of Michigan as an assistant professor in its law school. He resigned in 1948 to join the Seattle firm of Little, LeSourd, Palmer, Scott and Slemmons; and later became a partner in 1953. The firm engaged in general law practice in state and federal courts and with administrative agencies, specializing in tax and anti-trust matters.

In 1951, Cox was recalled to duty by the Army to serve as a member of the staff of the Army General School at Fort Riley, Kansas. There he was an instructor in intelligence training and later became administrative officer of the school's intelligence division. After that tour of duty, he rejoined his law firm in 1952.

In late 1955, he was named special counsel for the Senate Commerce Committee and, in that capacity he helped direct that committee's television inquiry of 1956-1957. He resumed his Seattle law practice in April 1957, but returned to Washington, D.C. for brief periods in 1958, 1959, and in 1960 to conduct additional hearings for the Senate Committee. In April of 1961, the Federal Communications Commission named him Chief of its Broadcast Bureau, and in 1963, he was appointed

FCC Commissioner by President John F. Kennedy.

Most commission observers, regardless of their opinion of his regulatory philosophy, generally agree that Cox was an uncommonly able Commissioner. Chairman Dean Burch, President Richard Nixon's choice to lead the Commission, called Cox, "A worthy and noble advocate of his position on the Commission. Although we disagreed frequently, it was not because of a lack of scholarship or candor."

Cox's appetite for work was awesome, and no item on a Commission agenda was too insignificant for him to prepare to discuss in detail. As a Commissioner, Cox was animated by a liberalism grounded his faith in the capacity and obligation of government to raise the public-interest quality of programming, by a" mom and pop" attitude toward broadcast ownership, and by a persistent skepticism of the willingness of broadcasters to operate in the public interest without close supervision. Cox said:

> I don't think the profit motive provides an incentive for the kind of programming the public needs. So long as they get an audience for what they do, broadcasters will do it, without regard to the needs of significant elements of the population that are not being served.

It was this attitude that propelled him, in 1962 , when he was still Chief of the Broadcast Bureau, into the center of a major controversy. Acting on the authority of the Commission, Cox instructed his staff to question renewal applicants whose proposed local live programming in prime time appeared to be inadequate. To many broadcasters, receiving the letters, the staff's questioning suggested that the inclusion of a proposal for sustained local live programming in prime time would speed Commission action on their renewal applications.

Commissioner Cox was a strong supporter of the Commission's proposal to loosen the networks' grip on prime-time programming, believing that the new rules would help stimulate new sources of programming. Cox also believed a multitude of editorial voices enabled the country's democratic system to function best, and strongly supported breaking up multimedia holdings within the same communities.

From March 1964 to November 1967, Cox served as Chairman of the Commission's Advisory Committee for Land Mobil Radio Services, and was also a member of its Telephone and Telegraph Committees. On September 1, 1970, Cox retired from the FCC after his seven- year term. The end of his time on the Commission became predictable when

Richard Nixon won the election in 1968.

Upon retiring, Cox joined Microwave Communications of America, Inc. as a senior vice-president, and worked in association as counsel with the Washington, D.C. law firm of Haley, Bader, and Potts. Working as counsel to this firm, Cox was engaged in the general practice of law as well as with specific activities in microwave transmission facilities for business.

By Gerald V. Flannery, Ph.D. and Ricky L. Jobe, M.S.

LOEVINGER, LEE
1963-1968

Lee Loevinger became President John F. Kennedy's fourth and final appointee to the Federal Communications Commission on June 11, 1963. He succeeded Newton N. Minow, who was Kennedy's first appointee in 1961.

Loevinger's appointment was a pleasant surprise to many followers of the FCC, because after three Kennedy appointees who were thought to be heavy regulators, Loevinger was viewed as a "hands-off" commissioner. He received unanimous endorsement from the committee that reviewed him and the Senate, which is responsible for approving presidential nominations to federal office.

Minow was an FCC commissioner who gave the appearance of wanting a large amount of influence in programming, but Loevinger was viewed as his antithesis by some. Minow challenged television programmers to change the "vast wasteland" that constituted television's airwaves. Broadcasting magazine reported that Loevinger, conversely, revealed to his interviewing committee that he would not wish to try to impose his tastes on the country any more than he would want anyone else to impose their tastes on him. He appealed to the committee and broadcasters because he had no desire to wield a heavy regulatory hand in programming. Loevinger believed that the stations themselves are in the best position to make programming decisions for their viewers. He said, "My feeling is that if I am to err, I would rather err on the side of restraint." That was contrary to New Frontier appointees Newton Minow, E.William Henry and Kenneth A. Cox.

Loevinger came to the FCC from his position in the Department of Justice, where he was an assistant Attorney General in charge of the Antitrust Division. He served in that capacity from March 16, 1961, to his FCC appointment.

Prior to his work in the federal government, Loevinger enjoyed an active law career in Minnesota. He graduated *summa cum laude* from the University of Minnesota with a B.A. in 1933, and got his law degree from the same university in 1936. At Minnesota he captained the debate team, edited the campus magazine, presided over the U.M. Board of Publications and served as editor of the law review. His college honors included Phi Beta Kappa, Delta Sigma Rho, the Forensic Medal, the Alumni Weekly Fold Medal, a citation as a representative Minnesotan and membership in the senior men's honor society, Iron Wedge.

Upon graduation, Loevinger became an associate of the law firm of Watson, Ess, Groner, Barnett & Whittaker in Kansas City, Missouri. THe following year he left them to become a trial attorney for the National Labor Relations Board, eventually serving as a regional attorney.

He remained with the federal government and, in 1941, joined the Antitrust Division of the Department of Justice as an attorney. He entered the military during World War II to serve on active duty with the Navy. He advanced in rank from Lieutenant junior grade to Lieutenant Commander.

After the war, Loevinger returned to private practice in Minnesota. He became a partner in the law firm of Larson, Loevinger and Lindquist in Minneapolis. He remained with that alliterative group until 1961, when he took a position as Assistant Attorney General for the Justice Department. During his stint in the private sector, he acted as Special Counsel to the United States Senate Small Business Committee on a part-time basis. In 1950, Loevinger married the former Ruth Howe of Glencoe, Minnesota. They had three children: Barbara, Eric, and Peter.

Loevinger's legal career, whether or not it prepared him for a spot as an FCC commissioner, led to his membership in the Minnesota and the American Bar Associations, the Federal Bar Association, the American Judicature Society, the American Association for the Advancement of Science, and Sigma Delta Chi, a professional journalism society. He was admitted to practice law in Minnesota, Missouri, and the United States Supreme Court, as well as in the U.S. Court of Appeals for the Eighth Circuit, the Treasury Department, and the Interstate Commerce Commission.

In addition to his law career, he served as a visiting professor of Jurisprudence at the University of Minnesota Law School, as a lecturer on law in the University of Minnesota Medical School in 1953-1960, and acted as Chairman for the Minnesota Atomic Development Problems Committee in 1957-1959. Loevinger also authored several books and articles in the fields of antitrust law, legal logic and jurisprudence. His writings included leading articles on the use of science and computers in law.

Loevinger came into the FCC office in a strained period for the regulatory agency. It had come under fire in the late 1950's and early

1960's (see John Doerfer and Richard A. Mack) for corruption and overly-political influences, so his appointment as a staunch man of the law was received favorably.

Loevinger later became a partner in the law firm of Hogan and Hartson in Washington, D.C.and also served as vice-president and director of Craig-Hollum Corporation.

By Joe Lynch, M.S. and Gerald V. Flannery, Ph.D.

WADSWORTH, JAMES J.
1965-1969

Some people are destined for greatness. Some people have to fight every step of the way to make it to the top. Others just plain have it in their blood. James Wadsworth was such a man. Wadsworth's great great uncle, Jeremiah Wadsworth, was a general under the command of George Washington. His great grandfather, James Wadsworth, commanded a division in the Civil War. His maternal grandfather, John Hay, was President Lincoln's private secretary and Secretary of State. His father was U.S. Senator James W. Wadsworth of New York.

Wadsworth was born on June 12, 1905 in Groveland, New York. He received his A.B. degree from Yale University in 1927 as well as L.L.D. degrees from Alfred University, Bowdoin College and Willmington College. Wadsworth was also involved with managing a farm that he received as a birthday gift. As the bloodline suggests, one can not take the country out of the boy.

Destiny caught up with Wadsworth when he embarked on a political career. In 1931, he began ten years of service as a New York State Assemblyman. He continued his involvement with farming and the state legislature until 1941 when he took the post of assistant manager of industrial relations at the Curtis-Wright Corporation in Buffalo, New York.

He remained there until 1945 when he was appointed Director of the Public Service Division of the War Assets Administration for a two year term. From 1946 to 1948, Wadsworth was the Director of the Government Affairs Department of the Air Transport Association of America. He left government service in 1948 to serve as special assistant to Paul G. Hoffman, head of the Economic Cooperation Administration. Wadsworth returned to the public service as the Director of the Civil Defense Office (1950) and moved on to become Deputy Administrator of the Federal Civil Defense Administration (1951) and then acting Director in 1952. During his tenure he was instrumental in drafting and implementing several civil defense plans.

President Dwight D. Eisenhower appointed Wadsworth as the

Deputy U.S. Representative to the United Nations in 1953. Wadsworth felt if the United Nations could be successful in negotiation then there would not be a need for civil defense. Wadsworth was active in the U.S. delegation that developed the International Atomic Energy Agency. From 1958 to 1960, he headed the U.S. delegation to the nuclear test ban conference in Geneva. Wadsworth helped negotiate a partial ban on nuclear weapons. For his efforts, Wadsworth received the Eleanor Roosevelt Peace Award in 1963 and the Contribution to Peace Award of the United World Federalists in 1964.

President Lyndon Johnson appointed Wadsworth to the Federal Communi-cations Commission on April 13, 1965. He began his term May 5th. Wadsworth was described as an "unpredictable maverick" by his peers. His philosophy toward communication regulation was "the less, the better". His actions reflected his reference to himself as a moderate to liberal Republican. During his tenure with the FCC, Wadsworth cast a key vote in the first case in which the FCC ever decided to deny renewal to a major television licensee. In another incident, Wadsworth voted to withhold renewal from a second licensee.

In 1966, the FCC saw the dawning of a new era in citizen participation in license renewal. In 1964, the FCC had renewed the license of WLBT, in Jackson Mississippi, despite claims by the United Church of Christ that the station was biased against blacks in employment and programming. At the time, Jackson had a 45% black population. The United State Appeals Court ruled in 1966, that the FCC should have taken the interests of the group into consideration. The Appeals Court decision opened the door for what the court called "responsible and representative" groups in FCC proceedings. A biracial group called Communications Improvement, Inc. took over the station and ran it successfully for eight years until a new biracial ownership group took over.

Two years before the end of his term expired, Wadsworth retired from the FCC. He resigned to join the American team negotiating a charter for the International Telecommunications Satellite Consortium known as Intelsat. A year later (1970), Wadsworth retired.

Wadsworth also served in various other positions. He was Chairman of the New York State Assembly Public Welfare Committee (1937-1941) and the New York State Joint Committee on Employment of Middle Aged (1939-1941); he was Chairman of the Buffalo Mayor's committee on Child Care (1944-1945), Chairman of the Board of trustees of Freedom House (1961), trustee of People to People, served on the U.S. Committee of the Dag Hammarskjold Foundation, and also served as president of the Peace Research Institute.

Drawing from his experiences, Wadsworth authored, <u>The Price of Peace</u>, <u>The Glass House</u>, and co-authored <u>A Warless World</u>. Wadsworth, an Episcopalian,died on March 14, 1984. He was survived by his wife, Mary A. Wadsworth, and one daughter.

By David A. Male, M.S. and Gerald V. Flannery, Ph.D.

HOUSER, THOMAS J.
1971-1971

Thomas J. Houser served seven months on the Federal Communications Commission. During his short term in office, many proposals were made, but very few completed before he retired from his position on October 6, 1971.

Houser, in 1966, was the campaign manager of Senator Charles Percy, a Republican from Illinois. After his task as campaign manager was completed, Houser practiced private law in Chicago. At the time of his appointment to the FCC, he had been Deputy Director of the Peace Corps for one year.

On January 5, 1971, President Richard Nixon appointed Commissioner Robert Wells and Commissioner Thomas Houser to the FCC. Both were given recess appointments and both were confirmed by the Senate on February 26, 1971. With these two appointments, the Commission was back to its full strength of seven members; also, the FCC started operating under a Republican majority for the first time since 1961. Houser was sworn in time for him to attend his first Commission meeting, an experience he later described as "intellectually stimulating."

Houser was named to the Commission to fill out the remainder of the term of Robert Wells, which expired on June 30,1970. He wound up serving a few months beyond that date. His main concern, when he accepted the appointment, was that he wouldn't be in office long enough to make a substantial contribution.

Houser was always a strong opponent of the use of labels like conservative and liberal on new appointees. He argued that new Commissioners should be given a chance to develop their views on issues confronting the agency.

As an attorney in Chicago, Houser had practiced before the Interstate Commerce Commission and state regulatory authorities as counsel for railroads and railroad associations. As a result of this experience, he had some notions about the difficult task of government regulation.

Houser felt strongly about simplifying the regulatory process. He explained that as a railroad lawyer he dealt with statutes having "a patchwork quality" because of amendments enacted over a long period of time, which reflected the differing views of the legislators involved. As a result, Houser felt that the industry being regulated had reason to feel harassed and at times confused by government regulation. To solve this conflict, Houser suggested drafting specific rather than broad regulation. Some government officials voiced concern about Houser's views about regulation feeling his opposition to strict regulation could hurt the agency's role. However, Houser felt strongly about violators, and advocated a policy of fines and revocation of licenses at the renewal time, to punish the offenders.

During Houser's time, the Commission reviewed its policies on license renewal time, procedures, and regulations; and also established strict criteria for determining which applications could be renewed routinely and which would require closer study.

Houser also proposed an incentive system for regulation enforcement. He suggested that the government provide the broadcast industry with tax incentives to create more public-interest programming without having to worry about any loss in profits. He also proposed the creation of an office of public counsel to help citizens challenge not only the broadcasting and telephone establishments, but the Commission's bureaucracy as well.

On September 22, 1971, towards the end of his term in office, the FCC faced serious charges by a Senate subcommittee. The subcommittee suggested that the agency had no expertise and little power to regulate the "multi-million dollar" advertising of drugs on television and radio. An FCC member suggested during a hearing that the Commission should ban all broadcast advertising of over-the-counter drugs for altering moods, mainly sedatives, sleep aids and stimulants. The subcommittee had previously heard expert testimony that the over-the-counter drugs for altering moods were generally ineffective.

Senator Gaylord Nelson of Wisconsin cited testimony indicating that widely advertised products were not ineffective, but could have adverse effects on some users. At the same hearing Commissioner Houser said that he agreed with Chairman Dean Burch that the FCC had neither the man-power, the expertise, nor the overall regulatory role to oversee drug advertising. He was quoted in The New York Times article in September 1971, as saying the television tube seemed to have become " a virtual electronic hypochondriac bringing home the theme for all of life's problems there's a pill."

Houser was particularly critical of drug advertisements, primarily of vitamins, that were aimed at children. He said three corporations, Brystol Myers, Miles Laboratories, and Sterling Drug, Inc., spent a total of $19 million advertising children's medicine on television in 1970. He felt children should not make medical decisions.

On October 6, 1971, he announced his retirement to resume the practice of law with the firm of Leibman, Williams, Bennett, Baird and Minow in Chicago.

By Marisol Ochoa Konczal, M.S. and Gerald V. Flannery, Ph.D.

WELLS, ROBERT
1969-1971

He wasn't a lawyer, but his knowledge of running a small radio station made Robert Wells a strong contributor to the Federal Communications Commission.

Robert Wells was born in Garden City, Kansas, on March 7, 1919. He was raised on a farm and attended Garden City High School. Later, while attending Garden City Junior College, Wells began his broadcasting career as an announcer with KIUL in Garden City. After working at the station from 1936 to 1939, he moved to Great Bend, Kansas, where he took another announcing job with KVGB.

Wells was inducted into military service in December 1940 and served with the 35th Infantry Division. He was an enlisted man until he was commissioned from military police officer candidate school in 1942. Wells did overseas duty in the Mediterranean during World War II and was released from active duty in December of 1945 as a Captain. He went on to serve as a Major in the Army Retired Reserve.

In 1946, following his discharge from the military, Wells returned to KVGB in Great Bend. He married Katherine Lovitt of Great Bend, Kansas in April 1947 and they had two sons, Kim, born in 1948 and Kent, born in 1954. Also in 1948, he moved back to his hometown of Garden City and assumed the position of station manager for KIUL, Inc. and, in addition to his work at the station, he was publisher of the Garden City Telegram from 1957 to 1961.

In 1961, Wells moved on up the corporate ladder when he became General Manager of the Harris Radio Group. For eight years Wells managed radio properties in the five markets where the chain had stations. He also became a minority stock holder in the operation.

President Richard Nixon announced his intention to nominate Wells as a member of the FCC on September 17, 1969. In mid-October of that year, Wells went before a Senate committee and, after stating that he had sold his interest in the Harris Radio Group, was informally assured Senate confirmation.

On November 6, 1969, Wells was sworn in by Chief Hearing Examiner Arthur A. Gladstone as a member of the FCC. Present at the brief meeting in the Commission meeting room were Dean Burch, who had been named the new FCC chairman five days earlier, and other members of the Commission and staff. Wells was appointed to complete the term of James J. Wadsworth, who left the Commission to join the U.S. delegation to the International Telecommunications Satellite Conference.

The nomination of Wells satisfied a common complaint among members of the communications industry that a broadcaster should be named to the FCC. It was felt that the broadcast experience that Wells had gained over the years would help bring some issues to a practical, real world level. On January 19,1970, Wells was named Defense Commissioner. He had proven himself to be a valuable member of the Commission and, on January 6, 1971, he was given a full term with the FCC.

As it turned out, Wells would not finish the new term. On October 21,1971, he announced that he was resigning. The letter of resignation indicated that he wanted to return to his hometown of Garden City, which he did. Wells also had political ambitions and was considering a bid for the Republican nominee in the Kansas gubernatorial race in the coming year.

Broadcasting magazine, in it 1971 chronology of important events, listed some of the things the agency dealt with in that time. Some of them were: the U.S. House and Senate agreed on legislation outlawing cigarette commercials on radio or television; the FCC banned major networks from controlling more than three hours of programming during prime time, creating access time; the White house created the Office of Telecommunications Policy; and the FCC began its investigation of commercials on children's programming.

Wells held offices in civic and professional organizations. He was president of both the Rotary Club and the JayCees, lay leader of his local Congregational Church; Chairman of the Kansas Forestry Fish and Game Commission; Kansas Council on Outdoor Recreation; two terms as president of the Kansas Association of Radio Broadcasters; and outstanding young man of the year for the State of Kansas.

An editorial in Broadcasting magazine on November 1, 1971, reflected the industry's feelings about the departing Commissioner In part it said:

> It is hard to conceive how conditions could have been worse for broadcasters during the past two years. Yet they would have been if Mr. Wells hadn't been there. He knew from experience what

station operations entailed and thus was often able to block some of the wildest thrusts of belligerent staff members and commissioners.

The FCC would miss Wells's first-hand knowledge of what it was like to run small-market station in the real world. Although not a lawyer, his common-sense judgements were not easily overridden. His performance more than justified the appointment of a broadcaster to the FCC.

By Gerald V. Flannery, Ph.D. and Richard E. Robinson, M.S.

JOHNSON, NICHOLAS
1966-1973

Nicholas Johnson was the youngest man ever to be appointed by a president to the Federal Communication Commission. But, being the youngest was not his only attribute. He was also the most controversial Commissioner in the thirty year history of the FCC. The mention of Johnson's name turned some broadcasters purple with rage, others lapsed into eloquent profanity, and still others released primal screams that found airwaves their own stations could not reach. Johnson's adversarial style shook up the maritime industry when he directed the U.S. Government's Maritime Administration, but his performance there was a minor tremor compared to the earthquake of controversy surrounding his actions as FCC Chairperson.

Nicholas Johnson was born on September 23, 1934, just three months after the creation of the FCC by Congress. Wendell Johnson , a supporter of the creation of the FCC and professor of speech at the University of Iowa, could not have envisioned a more perfect future for his son. Eighteen years later, in 1952, a young Nicholas Johnson married Karen Mary Chapman; immediately afterwards,he began his formal education at the University of Texas. By the time he finished law school with honors in 1956, he and had started a family. Eventually, they were the parents of Julie, Sherman, and Gregory Johnson.

He not only married and started his family at a young age, but also became somewhat of a prodigy in his career. In 1958, at the age of 24, he began as a clerk for Circuit Judge John R. Brown of the U.S. Court of Appeals. In 1959, he became a clerk with U.S. Supreme Court Justice Hugo L. Black. Like his father, he became a professor in 1960. He remained a law professor with the University of California, Berkeley until 1963; then, he became an associate member of the Washington, D.C. law firm of Covington & Burling. President Lyndon Johnson selected Johnson, the 29 year old stranger from Iowa, to be the Maritime Administrator in February 1964. Johnson stated in his book How to Talk Back to Your Television Set, "Without his (President Johnson's) commitment and confidence I would have never had been offered, and persuaded to accept, the position of FCC Commissioner." President Johnson appointed young Johnson to the Commission in 1966.

Johnson was everything the President wanted in an appointed Commissioner. He was young, bright, brash, and convinced that all of society's problems were social dysfunctions susceptible to systems analysis. Turbulent times for the communications industry were on the horizon. The President found a pro-regulation, courteous man who could respect the opinion of others, but still meet the challenges of balancing "public interest" with free speech and profitable broadcasting. President Johnson also appointed Rosel Hyde as Chairman for a full seven year term. The broadcasting industry was optimistic about the two appointments, especially Hyde who was known for his stance that broadcasting would improve if the governmental climate was favorable.

Media optimism quickly vanished. Less than two years after his appointment, Johnson became the most criticized Commissioner by the broadcasting industry. At various times, Broadcasting magazine referred to young Johnson as "self-appointed savior," "consummate nuisance," "arrogant," "noisiest dissenter," and even "neophyte." Johnson's unpopularity was a direct result of his belief in strict regulation of the broadcast industry. His hard-line beliefs were no surprise to an administration who was warned that his "public trust" stance could lead to strict regulation and governmental control of the industry. The decisions he made in cases as FCC commissioner only reflected his hard-line personal beliefs.

The   International Telephone and Telegraph and American Broadcasting Company (ITT / ABC) merger case was decided in June 1967. Hyde, Chairman at the time, voted to allow a merger between ITT / ABC . Johnson, disagreeing with Hyde's view, was among the dissenting votes. ITT was a sprawling international conglomerate of 433 separate boards of directors, that derived 60 percent of its income from holdings in at least forty different countries. It was the ninth largest industrial corporation in the world. Half of ITT's U.S. income came from government defense and space contracts. ABC was one of the major communication networks in the world. ABC, at the time had interest in, and affiliations with, the "Worldvision Group." Johnson believed that the ITT / ABC merger would allow ITT to use the network as part of its public relations, advertising, or political activities. Johnson used the following reasoning for his dissent, "To do an honest and impartial job of reporting the news is difficult enough for the most independent and conscientious of newsmen, but the mere awareness of the ITT interest would make reporting news and factual documentaries objectively an impossible task." Johnson was not alone in his belief. During the merger process,

ITT stated how they planed to stay out of ABC's newsroom but, at the same time ITT would try to influence newsmen and editors to give them favorable coverage during the merger process. After testimony from reporters was given, and court cases were filed, ITT aborted the merger plan. The Commission's action was pending before the U.S.Court of Appeals, but no judgment was made because of ITT's withdrawal on New Year's Day, 1968.

ITT / ABC was not the only controversial decision of Johnson's term. At Hyde's request, CATV regulation was under consideration by the FCC. The FCC's question then was to consider how they might mold CATV into an instrument that could compete equally and not eliminate free broadcast operations. Johnson again advocated strict regulation, even with new media like cable. In Johnson's book, he mentioned reasons why he wanted strict regulations imposed on cable. He believed that cable was not available to most citizens especially poor or rural citizens. It did not offer free services, but several channels to those who could afford cable. He believed that the "public interest" would be best served by regulating CATV because, "if moves are not made very soon to channel the future growth of CATV along lines responsive to social needs, it will likely be too late."

Through the cable question many more issues arose that needed attention. Just like free broadcasting, some questions of ownership, availability, and market distribution existed. Cable would increase lobbying against the home satellite communication systems currently fighting for existence   Johnson was in favor of home satellite communication because it would free the public from the strict control of the networks and allow more socially responsible programming. He was visionary in one aspect of cable. Johnson weighed free TV versus pay CATV as if they could exist in the same market equally. Others on the Commission could not see coexistence as a future.

Johnson considered television a menace to society. In his article, "What Can You Do to Improve TV," which appeared in TV Guide, he recommended strict self regulation by the viewers of broadcasting networks. He believed that public interest groups should get more actively involved in television and radio renewal licensing. He advocated the reduction of violence and the increase of minority programming. He wanted CATV to carry out the social responsibility that the networks and other already established broadcasters did not. As an example of reform he wanted to see, Johnson provided in the TV Guide article the following scenario:

Each of the three networks could be required to provide a single hour of prime time programming each evening to public service programming between 7 and 10 p.m.... thus viewers would have the choice of something other than advertising to watch between breaks in other networks programming.

That scenario provides the framework of logic Johnson was constantly using to argue for more socially responsible programming. He also advocated no commercial programming during news broadcasts, and regulating the size of news staffs, budget and income. His idea of regulating news broadcasts was often represented as the act of an evil prophet sent to regulate the press and free speech. The media interpreted his remarks to be anti-free speech instead of being merely a listing of what he believed to be wrong. Wrong for Nicholas Johnson as a Commissioner was the profitable exploitation of speech in the mask of public champion and freedom.

Johnson's controversial beliefs and actions were constantly attributed to his youth and arrogance but it was not his youth that created controversy, it was his ideology. In his private life, he was a democrat and practiced the religion of Unitarian Universalist. The royalties from his published works, while he was Commissioner, were contributed to organizations devoted to improving television's contribution to the quality of American life. He believed in the FCC's mission to serve the" public interest" and carried out that mission to the best of his ability. He was a man who held nothing back.

By Dawn Vogelsang, M.S. and Gerald V. Flannery, Ph.D.

LEE, ROBERT E.
1953-1981

Robert E. Lee was born in Chicago on March 31, 1912. He graduated from DePaul University College of Commerce and Law, was married twice and had three children. Lee, a Republican, was appointed to the Federal Communications Commission by President Dwight D. Eisenhower on October 5, 1953, for his first seven year term. Members of the Senate were concerned about Lee's affiliations with Senator Joseph McCarthy and Texas oil mogul H.L. Hunt, plus, Lee had little experience in broadcasting. Despite that, Lee was confirmed by the Senate 58-25.

Prior to the FCC, Lee served as an auditor, a Special Agent, and an Administrative Assistant for the FBI under J. Edgar Hoover. In addition, he served as Director of Surveys and Investigation for the Appropriations Committee of the U.S. House.

Lee, speaking to the New England Chapter of American Women, made his philosophical position clear, saying there should be a minimum of government interference in broadcasting. During his first term, Lee developed an interest in Ultra High Frequency television and also called attention to the slow development of ETV. Lee felt that UHF would expand programming by opening up seventy new channels but UHF had a slow beginning, until 1964, when Congress adopted a law, requested by the FCC, requiring all new television sets be equipped to receive UHF.

Lee, a staunch advocate of free speech, recognized that some legal safeguards were needed to protect the public from profanity and subversive comments but felt that an industry- wide code of ethics would be far superior to government regulation. In 1954, a bill went before Congress that would prohibit ownership of radio stations by newspapers in cities with populations of 100,000 or more. Lee opposed it feeling the community would benefit from radio and newspaper services working together.

Lee became embroiled in controversy again in 1958 when some

commercial broadcasters were suspected of providing favors to FCC employees in return for special consideration. A House subcommittee questioned Lee about his travel expenses. He provided an accounting, saying he resented the implication he could be "bought." The following year gave rise to the "payola" scandal in broadcasting and Lee warned such practices were in violation of the Communications Act and could lead to loss of license. On June 23,1960, Lee was confirmed for a second term.

The launching of the first active communications satellite on July 10, 1962, was a technological harbinger for things to come and Lee came to the regulatory forefront in May of 1965 when he voted against allowing the Communications Satellite Corporation the sole right to build and operate ground stations to transmit / receive signals to and from space satellites for a two year period. His stand on regulation was tested again in March 1965 when the FCC adopted a rule to decrease network influence on television programming by limiting the amount of prime time that could be controlled by the three major networks. Lee felt that the ruling would give advertisers too much control over the programming process and offset the balance of control.

Near the end of his second term, Lee focused his attention on radio, proposing that the over 4,000 radio stations should merge their management systems in order to coordinate the industry as a whole. It didn't fly. In addition, in March 1966, Lee predicted that FM radio would surpass the popularity of AM in the next ten years. It did.

In July of 1967, Lee announced that he was retiring to take a job with a corporation for the same $27,500 he was receiving as FCC Commissioner. However, President Johnson appointed him for a third term in August and Lee accepted.

In March 1971, the FCC ruled on data processing saying all large communications companies, such as RCA and AT&T, had to set up totally separate affiliates in order to sell data processing services. Three Commissioners voted against that ruling, claiming that denying the common carrier access to its own computer services affiliate was an example of regulatory overkill.

During April 1971, the FCC reported an increasing number of complaints that films and broadcast programs were degrading to ethnic and racial groups. Lee reported that in 1970 over 130 complaints were received. He recognized offensive portrayals were protected under the First Amendment but told the House Sub-committee that he hoped the industry would use proper judgement in self regulation.

The Christmas season of 1971 was marked by an FCC withdrawal from an ongoing investigation into AT&T rates in the planning stage since

1965. In November of 1972, Lee would vote for an increase in rates, with the understanding that the increase be applied to business hour calls and operator assisted calls. That change brought the total increase in the 1971-1972 year to $395 million. In an unrelated matter, the FCC was also investigating charges that AT&T was discriminatory in its employment practices.

In February 1972, the FCC adopted a policy to limit the growth of cable television in the big cities in order to stimulate cable television growth in the smaller ones. Lee voted against it because formal adoption of the policy without public hearing was, in Lee's opinion a violation of the Administrative Procedure Act. In ten short years, communications satellite technology had progressed to the point where it was accessible to industry. In June 1972, the FCC voted to permit all qualified applicants to provide satellite service to transmit television, telephone, telegraph and computer data signals.

Lee was reappointed for his third term by President Richard Nixon on May 18, 1974. In September, the FCC denied renewal to eight Alabama ETV stations because of a history of discrimination against blacks in hiring and programming. That decision sent notice to citizen groups that something could be and was being done about discriminatory practices. Lee and Charlotte T. Reid voted for renewal of the licenses but lost in a 4-2 vote.

On September 26, 1975, the FCC reversed the policy requiring equal time on radio and television broadcasts of news conferences and political debates for candidates. Lee cast a dissenting vote feeling that action would give the broadcaster authority to decide what is newsworthy and to offset the balanced needed to assure a fair election. On May 7, 1977, the FCC ruled 4-3 that broadcast stations were not required to sell a political candidate the specific amount of time requested but that the candidate should have the same right to air time as a commercial advertiser, and should be charged the lowest commercial rate. Once again Lee dissented.

On April 13, 1981, Lee was named Acting Chairman of the FCC, but he only stayed until June 30 of that same year. During his short tenure as Acting Chairman, Lee endorsed an FCC policy allowing direct broadcasting from satellites to homes. In August of that year, the Satellite Broadcasting Company elected the former FCC Commissioner / Chairman to its board. In December, Lee was named Vice-President of Corporate Communication for Ricoh Corporation.

By David A. Male, M.S. and Gerald V. Flannery, Ph.D.

BURCH, DEAN
1969-1974

It was not long after Dean Burch became Chairman of the Federal Communications Commission that he let his overall administrative policy be known. In December 1969, just five weeks after being sworn in as Chairman, Burch gave a speech in which he denounced the idea that the quality of radio and television programming could be improved through dictates from the nation's capitol.

He also said he was not enamored with the idea of breaking up multi-media combinations or setting up standards for the amount of news, public affairs, or other types of programming radio and television stations broadcast. " I am not impressed with the idea that the FCC is better able to handle programming than somebody whose job it is," he said. Here indeed was a conservative Republican FCC Chairman many of the nation's broadcasters could learn to like.

Burch's felt that government should leave the broadcasting industry alone, as much as possible, so that it might police itself in free market place. He did, however, pinpoint two areas he would like to see closely watched by the FCC : the upgrading of children's programming and regulation of commercials in number and kind.

Burch became Chairman of the FCC on October 31, 1969, replacing retiring Chairman Rosel H.Hyde and becoming the fifteenth Chairman of the FCC up until that time. He was nominated to the Commission by President Richard Nixon on September 16, 1969, and confirmed by the U.S. Senate on October 30. A resident of Arizona, Burch was a member of the law firm of Dunsheath, Stubbs and Burch in Tucson when named to the Commission.

Burch's confirmation was not to come without controversy. At a meeting of the Senate Communications Subcommittee in October 1969, Burch was called a "rich, white racist" by Absalom Jordan, Junior, national chairman of Black Efforts for Soul In Television (BEST), which called for the appointment of blacks to the FCC. In responding, Burch said he was neither rich nor a racist, saying " I feel sorry for Mr. Jordan for making such a blanket condemnation without further evidence."

He was born in Enid, Oklahoma on December 30,1927, the son of Bert Alexander and Leola (Atkisson) Burch. His father was a guard with the Federal Bureau of Prisons, and as the result of changing duty assignments, Burch attended elementary schools in Norman, Oklahoma, Leavenworth, Kansas and San Francisco, California. The family lived on Alcatraz Island from 1940 to 1948.

Following graduation from high school in San Francisco, Burch ,in 1946, enlisted in the U.S. Army. He attended Officers Training School in Fort Knox, Kentucky and served with the 7th Calvary Regiment in Tokyo. He left active duty in 1948 and achieved the rank of Colonel in the Judge Advocate General's Corps Reserve.

Burch was a graduate of the University of Arizona and received his law degree there in 1953; he was a member of Phi Delta Theta and Blue Key. Following his admission to the bar in 1953, he was appointed assistant to the Attorney General of Arizona. He first went to Washington, D.C. in 1955 as legislative and then administrative assistant to Senator Barry Goldwater, then returned to private practice with Dunseath, Stubbs, and Burch from 1959 to 1963. He was deputy director of the Goldwater for President Committee, 1963-64.

He was Chairman of the Republican National Committee from July 1964 to April 1965. In January, Governor Jack Williams named him to the Arizona Board of Regents. He resumed work as a partner with Dunseath, Stubbs and Burch from 1965-1969.

Broadcasting magazine, in it 1971 chronology of important events, listed some of the things the agency dealt with in that time. Some of them were: Vice President Spiro Agnew charged the networks with biased reporting; the U.S. House and Senate agreed on legislation outlawing cigarette commercials on radio or television; the FCC banned major networks from controlling more than three hours of programming during prime time, creating access time; the White house created the Office of Telecommunications Policy; the FCC began its investigation of commercials on children's programming; the FTC urged the FCC to expand its Fairness Doctrine to cover commercials; and the FCC rules to allow multiple entry into satellite systems.

Burch's term as Commissioner was to run to June 30,1976, but he resigned on March 8, 1974 and became counselor to President Richard Nixon from March to August 1974, and to President Gerald Ford from 1974 to 1975. He was the senior advisor to the Reagan-Bush Committee from July to November 1980 and again from August to December 1984. During the period of November 1980 to January 1981, he was Chief of Staff to Vice President George Bush. He also had been head of the U.S.

delegation to the First Session of World Administration Radio Conference on Use of Geostationary-Satellite Orbit & Planning of Space Service in Geneva, Switzerland.

Burch became a partner in the law firm of Pierson, Ball & Dowd in Washington,D.C. in 1975, specializing in communications law. Burch married the former Patricia Meeks on July 7, 1961. They had three children, Shelly, Dean A. and Dianne. He listed his hobbies as golf and tennis.

By Michael A. Konczal, M.S. and Gerald V. Flannery, Ph.D.

REID, CHARLOTTE T.
1971-1976

The only child of Edward Charles and Ethel Stith Thompson, Charlotte Reid was born in Kankee, Illinois on September 27, 1913. The Thompson family eventually moved to nearby Aurora, where Reid attended public schools. Like her parents, she was musically talented and took singing lessons from a local voice teacher for several years. After her graduation from East Aurora High School in 1930, she enrolled in the teacher education program at Illinois College in Jacksonville. In her spare time, she continued to study voice. Determined to devote herself to a singing career, she left Illinois College in 1932 without taking a degree. For a time, she studied under Louise Gilbert, a noted Chicago voice coach, who gave her free lessons because she was a promising student. By making a number of singing commercials and by singing for little or no money on local radio stations, she gradually gained some exposure. Four years later, she was chosen from more than 100 vocalists to be featured on Don McNeil's Breakfast Club. She sang regularly on the program until shortly after her marriage to Frank R. Reid, Jr., an Aurora attorney in 1938. They had four children..

Reid worked closely with her husband when he ran for nomination to Congress in the 1962 Republican primary. He died shortly after he was named Nominee to the House of Representatives from the Illinois Fifteenth Congressional District. The GOP county chairmen in the district selected his widow as the party's candidate although her political experience was minimal. Reid polled 63.3% of the votes cast, and she was elected to Congress in November 1962. She was sworn in on January 9, 1963. She was re-elected four times after that. When elected, she was one of 12 women in the House of Representatives and the only freshman Congresswoman.

During her years as Congresswoman, she served on many committees:

The House Committee on Interior and Insular Affairs (1963-1967)
The Committee on Public Works (1965-1967)
The House Committee on Appropriations (1967)

House Committee on Appropriations' subcommittees: Foreign Operations and Labor Health, Education and Welfare
The House Republican Policy Committee (1963-1965)
The House Committee on Standards of Official Conduct (1970)

Reid also worked in support of the public funding of the National Cultural Center, now the John F. Kennedy Center for Performing Arts. Her memberships also included:
The Board of Trustees of the Federal Woman's Award
Business and Professional Woman's Club
The Altrusa Club
Honorary member of Gamma Beta Sorority
She held an honorary degree of Doctor of Laws from Marshall Law School, Chicago and one from Illinois College.

On July 2, 1971, President Richard Nixon named Reid to succeed Thomas J. Houser on the Federal Communications Commission. She was confirmed by the Senate on July 29 to a seven year term on the FCC. Nixon asked Houser to remain on the Commission until October 1, 1971 because the administration needed Reid's support on several important measures before the House.

Charlotte T. Reid was the first woman to become a member of the FCC in more than twenty years. The only woman to precede her was Frieda Barkin Hennock (1948-1955). Reid was a legislator for nine years prior to her appointment to the FCC and she had some experience in broadcasting from her singing career to provide her with a good working knowledge of the Commission and its proceedings. It is ironic that she would be chosen for a position with the FCC because of her dislike for television. She said she only had time for the news and special events. She was a believer in media self-regulation, and once made the comment that "neither broadcasters nor advertisers want seven people in Washington to be arbiters of public communication." In her opinion, the FCC was not designed to act as national censor. "What is acceptable in California may offend the Midwest," she explained in an interview for the Akron Beacon Journal, reprinted in part in Variety. "I turn off programs which offend me."

Reid, as Commissioner, was interested in the Fairness Doctrine. She favored restricting the equal time provision to elections for national office and opposed the prime time access rule which returned the lucrative 7:30-8:00 p.m. (EST) time slot to the local stations. Prior to her activities as Commissioner, it was believed that she would support Burch, who was Chairman at this time, more than Houser did. She was an extreme conservative, yet she believed that control of the broadcasting industry should remain with the people at home: Government should have little control.

Broadcasting magazine, in its latest chronology of important events, listed things the FCC did in that period. Some of them were: the White House moved to eliminate the "Fairness Doctrine while the FTC asked the FCC to expand it to include commercials; the FCC developed a package of cable television rules; the FCC began broadcast deregulation by dropping seven technical rules; the Senate began hearings on television violence; the FCC ordered a ban on future cross-ownership acquisitions of newspapers, radio and television; and the FCC began examining it radio rules with an eye to further deregulation.

By Gerald V. Flannery, Ph.D. and Carla P. Coffman, M.S.

WILEY, RICHARD E.
1971-1977

Richard Emerson Wiley was born in Peoria, Illinois, on July 20, 1934. The son of Joseph Henry and Jean Wiley, he was raised in Winnetka, Illinois, an affluent suburb of Chicago. Wiley attended New Trier High School, one of the best public schools in the state of Illinois. After graduating from there in 1951, Wiley enrolled at the Northwestern University near Evanston. In 1955, he obtained a Bachelor of Science degree, graduating with honors. In 1958, he obtained his J.D. degree from the Northwestern University School of Law. From 1959 until 1962, he worked in the United States Army Judge Advocate General's Office at the Pentagon, where he attained the rank of captain. While he held that post, Wiley attended Georgetown University Law School as a part-time student. Then, in 1962, his diligence and determination paid off and he was awarded an LL.B. degree. In 1963, Wiley decided to return to Chicago where he joined the prestigious corporate law firm of Chadwell, Keck, Kayser, Ruggles & McLaren. He remained there for six years, specializing in antitrust cases.

In 1968, he joined Bell & Howell as an assistant corporation counsel. Active in politics since his undergraduate days, Wiley took a three-month leave from his new post to direct state political organizations for Richard Nixon's presidential campaign.

In addition to pursuing his career as a practicing attorney, Wiley taught at the John Marshall Law School in Chicago from 1963 to 1970. While at the law school, Wiley founded and edited Law Notes, a legal quarterly. An active member of the American Bar Association, Wiley served on its special commission on campus government and student dissent; at one point he chaired its young lawyers section.

In 1970, Wiley decided to go into private practice and established the firm of Burditt, Calkins & Wiley with two colleagues. A few months later, however, he was appointed general counsel to the Federal Communications Commission as Henry Geller's successor. In response to repeated criticism that he was unqualified for a post in communications law, Wiley once said that the Commission needed diversity of thought. And if someone spent their whole career in one place, he'd develop attitudes; but f you'd been on the outside, you'd bring a different point of view.

After spending several months on the job, Wiley earned a reputation as a well-informed middle-of-the-roader who patiently heard both sides in a dispute. More conservative than his active, regulatory-minded staff, Wiley favored easing regulations and restrictions on the broadcasters. He objected to the court-ordered extension of the Fairness Doctrine to commercial ad spots. He also questioned the wisdom of a FCC proposal to break up multi-media holdings in an individual market.

On Wells' resignation from the FCC, effective November 1, 1971, President Richard Nixon appointed Wiley to the Commission to complete Wells' term, which was to end in June 1977. Since Congress was not in session at the time of Wiley's appointment, he agreed to serve as a recess appointee until confirmed by the Senate. Confirmed in May 1972, Wiley headed several special units within the FCC, including those dealing with the re-evaluation of the Fairness Doctrine and the possible deregulation of obsolete FCC rules.

In February 1974, Wiley succeeded Dean Burch, (who had been appointed counselor to Nixon) as Chairman of the FCC. Wiley was credited with raising the efficiency of the seven-member agency. A hard-working and ambitious Chairman, Wiley favored minimal governmental interference in the communications industry, preferring the competition of the marketplace rather than strictly enforced federal regulations. Wiley devoted most of his time to broadcast regulation, despite the fact that the FCC oversees the production / operation of a wide range of communications equipment, including electronic heart devices, domestic satellites, and the like.

An ally of broadcasters, Wiley chose as his first public forum the 52nd annual convention of the National Association of Broadcasters in Houston, Texas in March 1974. Eager to "break down the wall of mutual suspicion, distrust and fear" between the FCC and broadcast licensees, he reassured his listeners that he recognized the "changing, diverse and pluralistic" nature of the communications industry and was wiling to modify the FCC's position to reflect conditions within the industry.

Often criticized for its complicated adjudicatory procedures, the FCC in 1974 was constantly evaluated by the government. He replied to the concerns with the suggestion of a regular, exhaustive review of FCC procedures and policies to weed out some regulations and to simplify some rules.

Wiley instituted regular regional meetings of the FCC and encouraged free and open exchanges between the regional bureaus and the Washington office. Following guidelines established by the

Republican administrations of Nixon and Gerald Ford, Wiley fought increased government regulation of communications.

Under his direction, the FCC ruled to allow competition in the supply of communications terminal equipment and in specialized private-line services to give the American consumer unrestricted access to new communications technology. Wiley also initiated a plan for the reduction of the federal regulation of commercial radio stations in major markets.Wiley repeatedly criticized the time-consuming process of license renewal every three years. He argued that no industry could prosper in an atmosphere of "constant upheaval."

To provide a reasonable assurance of continuity and stability to the broadcasting industry, Wiley recommended simplifying the renewal procedures and extending the life of broadcasting licenses to five years. He further suggested that the FCC limit itself to an evaluation of a licensee's past public service performance, rather than concentrating an net profits and possible multi-media interests.

Although Wiley's term in office was scheduled to end on June 30, 1977, it was not until September 12 that President Jimmy Carter nominated Charles Ferris as Wiley's successor in the Chairmanship.

Broadcasting magazine, in its latest chronology of important events, listed things the FCC did in that period. Some of them were: the White House moved to eliminate the Fairness Doctrine while the FTC asked the FCC to expand it to include commercials; the FCC developed a package of cable television rules; the FCC began broadcast deregulation by dropping seven technical rules; the Senate began hearings on television violence; the FCC ordered a ban on future cross-ownership acquisitions of newspapers, radio and television; the FCC began examining it radio rules with an eye to further deregulation and later dropped the 1941 rule package;and the FCC began investigating the power of the three networks in the marketplace.

Richard E. Wiley lived in Arlington, Virginia with his wife Elizabeth Jean Wiley, whom he married in 1960. They had three children Douglas, Pamela and Kimberly.

By Gerald V. Flannery, Ph.D. and James Nunez, M.S. candidate.

HOOKS, BENJAMIN L.
1972-1977

The first black Commissioner on the Federal Communications Commission was Benjamin L. Hooks, a former judge and later head of the National Association For The Advancement of Colored People. He was sworn in by Judge William B. Bryant on July 5, 1972 and, what set him off, at first, from the other Commissioners was his race, but soon his personality and flair, coupled with his concern for minorities and the poor, set him apart.

Hooks was born in Memphis, Tennessee on January 31, 1925 to Robert Britton Hooks and Bessie White. He attended Lemoyne College for two years and then Howard University, leaving it in the early forties as World War II came along and he spent three years in the Army (1943-46), serving finally as a Sergeant in the 92nd Infantry Division. After the war, he went to Depaul University in Chicago and later earned his Juris Doctorate in 1948. He gained admission to the Tennessee bar the following year and practiced law in Memphis from 1949 to 1965. It was there that he met and married Frances Dancy. They had one daughter and two grandsons.

Hooks was appointed assistant Public Defender in Memphis in 1961 and served there for the next three years. In 1965, the Governor of Tennessee appointed him Judge of Division IV of the Shelby County Criminal Court; he ran for reelection to that post the following year and was elected to an eight year term. He only stayed in that judgeship for two years, resigning in 1968 to return to private practice.

Hooks had a varied career on his way to prominence. He combined the law and the ministry as early as 1956 when he was ordained to preach by the Baptist church. The Middle Baptist Church in Memphis appointed him pastor that same year, and eight years later he became pastor of the Greater New Mount Moriah Baptist Church in Detroit, Michigan. During these years as a lawyer and a pastor, he also became an active businessman, serving as vice president and co-founder of the Mutual Federal Savings and Loan Association of Memphis, from 1955 to 1969. He was also a member of the Board of

Directors of the Savings and Loan Association, and of the Tri-State Bank in Memphis until 1972. He also produced and hosted his own television program "Forty Percent Speaks," was a panelist on the television program "What Is Your Faith?" and worked as counsel for the Progressive National Baptist Convention.

Hooks involvement with the NAACP began in 1965 when he served as attorney for the organization for one year, but he would later return to the NAACP, this time as its director, in July 1977. He was a lifetime member of the NAACP, was on the Board of Directors of the Southern Christian Leadership Conference, the Tennessee Council on Human Relations, the Memphis and Shelby County Human Relations Committee; he also was a member of the American Bar Association, the Tennessee Bar Association, the National Bar Association, and the Judicial Council of the National Bar Association.

Hooks' career as a Federal Communications Commissioner did change the organization. Pluria Marshall, head of the National Black Media Coalition, speaking of him as a member of the Commission , said, "It was a big step in the right direction... We could always call on Ben to discuss things. We're very proud of his service."

Vincent Wasilewski, president of the National Association of Broadcasters, credited Hooks with opening his eyes to minority problems, saying "he sensitized the Commission and the industry." FCC Chairman Richard E. Wiley echoed those sentiments saying " he heightened the awareness of us all." Because of that, some considered Hooks as a "one issue" commissioner, but it might be more correct to say he regarded the minorities and the poor as his constituency.

The record shows that Hooks favored keeping the Fairness Doctrine and the Equal Time rule seeing them as avenues that the poor and the disadvantaged could use to gain access to the media. He recognized that his appointment was owed, in part, to the efforts of activists like William Wright, then head of the Black Efforts for Soul in TV, and of its predecessor, the National Black Media Coalition; and the work of many other blacks, to encourage the appointment of a black Commissioner. Nevertheless, he served notice of his independence from that support, when he backed the Commission's stance on the new Equal Employment Opportunity rules indicating that he saw more good in the policy than bad.

Hooks, in 1972, wrote a concurring opinion in a comparative hearing case arguing that license applicants, who included minority members among their ownership, should get preferential treatment to assure greater diversity of ownership. That view became part of the court's finding and led to the Commission reversing its previous decision.

In May of that same year, he persuaded the Commission to ignore precedent in the case of WLTH (AM), in Gary, Indiana, and allow it to renew its license so that it could be sold to a group of black owners. That action gave Gary, a city with a large minority population, its first black-owned broadcast station.

Broadcasting magazine, in its latest chronology of important events, lists several other things he dealt with during his tenure. They were: the FCC initiated an inquiry into the prime time "access rule;" the FCC took the firsts steps in deregulating and re-regulation; it set up two Equal Employment offices to work with Commissioner Hooks; the FCC began trying to control "topless" radio and suggestive material on TV; the National Black Radio network got underway with 41 stations; the FCC discussed whether to strip the Alabama Educational Television Commission of its nine educational television stations charging racial discrimination in programs and hiring; the FCC changed its policy on Equal Time; and it changed its radio rules.

Hooks left the FCC in July 1977 to answer what he saw as a "call" to head the NAACP as its Executive Director, allowing him to deal with a larger range of issues affecting blacks in America. He used his experience as a lawyer, judge, and FCC Commissioner, to vigorously advance the causes the NAACP was interested in, no longer constrained by the First Amendment considerations he had to deal with as a Commissioner. It is hard to measure exactly Hooks' effect on broadcast regulation but it is clear he had an impact on the Commission's dealing in EEO matters, in minority ownership, and sensitivity to the poor and powerless. His presence was felt.

Portions of this article appeared in FEEDBACK, The Journal of Broadcast Education Association.

By Gerald V. Hannery, Ph.D. and Bobbie DeCuir, M.S.

QUELLO, JAMES H.
1974-

President George Bush reappointed James H. Quello to the Federal Commuications Commission at age 77, extending his term to June 30, 1996. The energetic Michigan Democrat was happy to remain a Commissioner another five years, championing the forces of free market over federal legislation.

"I believe preservation and enhancement of the all-important free, universal, over-the-air broadcast service is the mainspring of American mass communication," Quello told <u>Broadcasting</u> magazine in 1991.

In the same interview, Quello expressed hopes that President George Bush would save him a seat on the FCC. That appointment did materialize, and with a record of two decades on the Commission, Quello's influence was still felt.

Quello had settled into radio in 1947 as promotion manager of WJR in Detroit. He moved up the station's hierarchy and by 1960 was vice president and general manager. His career led next to Capital Cities Broadcasting Corporation, where he served as vice president until retirement in 1972. The same steadfast enthusiasm he was later to exhibit on the FCC was evident in the longstanding service Quello paid as a member of the Detroit Housing and Urban Renewal Commission (1951-73), and the Michigan Veterans Trust Fund (19541-74). In addition, he held posts on the Michigan Governor's Special Commission on Urban Problems, the Governor's Special Study Committee on Legislative Compensation, United Foundation, Boys Scouts of America, and the American Negro Emancipation Centennial Board.

As a broadcast manager, Quello developed an enduring respect for the influence of the audience and the advertiser. When the tide of broadcast regulation began to ebb in the 1970s, Quello was an enthusiastic advocate of the FCC's new hands-off policy. When asked, he expressed a personal distaste for the increasingly violent and sexual content of much mass media.

Quello could identify no regulatory remedy for the problem, however. Instead, Quello called upon broadcasters to voluntarily temper their programming, and reiterated his hope that sponsors and consumers would use their influence to sway professional opinion.

Quello's concern over media fallout was substantiated by a rising public interest in television. Three months after Quello was sworn in as FCC Commissioner, the Watergate impeachment hearings were broadcast on radio and TV, followed in months to come by the presidential resignation ceremonies and the swearing in of President Gerald Ford. The number of cable television systems in the United States had grown from 70 in 1950 to 3,158 in 1974, according to the New York Times Encyclopedia of Television. The growth of armchair audiences foreshadowed greater increases ahead. In 1977, the largest television audience ever, an estimated 80 million, would tune in the "Roots" miniseries on ABC.

The president whose resignation was broadcast as Quello assumed his FCC duties was the very one who had nominated him just months before. Nixon's approbation, however did not guarantee a smooth ride through the Senate, where the question of Quello's appointment was debated for eight days. The legislators were apparently wary of this former broadcaster whose name incited such furor among special interest groups. For, although Quello's policies kept him within the warm regard of industry professionals, citizen groups were less appreciative of his laissez-fair outlook. Quello's was the longest Senate debate ever over an FCC appointment, and even once approved, confirmation was stalled for seven months.

During Quello's early years on the Commission, the FCC abolished most chain broadcasting regulations for radio. At the same time, the FCC began to actively encourage competition between broadcast and cable, lifting many limitations on the latter.

A power struggle between the FCC and the U.S. District Court of Appeals had begun six years before Quello took office. It continued seven more. The FCC, who typically grants routine approval to changes in radio programming, found its role being redefined by the court. Citizen's groups -- unhappy with the FCC's rubber stamp on format changes they opposed -- began securing court orders for FCC public hearings. Indignant FCC Commissioners (including Quello) called the detailed reviews ordered by the court an "unconstitutional" check on free speech, and "administratively a fearful and comprehensive nightmare." In what many thought a gutsy move, the FCC voted to disobey court orders, and instead staged a public hearing to examine for itself what role the

Commission should play in format review. The U.S. Supreme Court finally resolved the dispute in 1981, ruling in the FCC's favor that the Commission had no obligation to regulate radio programming.

By the time Quello was nominated for reappointment by President Ronald Reagan in 1984, Broadcasting magazine called the subsequent Senate hearing "as soft as ice cream of a hot summer day...a noncontroversial nomination." Consumer protests, it seems, were quenched not by a change in Quello's policy, but by his diplomacy. The Commissioner's repeated vows to keep "an open door policy" and to "examine both sides" blunted the barbs of consumer criticism. Quello was not deaf to the complaints of consumers. He directed them, rather, on a less circuitous route directly to the broadcasters themselves.

All American corporations exist by the will of the people, Quello told Broadcasting magazine in 1984.

> It behooves all corporations, acting in their own self interests, to conduct themselves with a keen sense of social purpose, not only economic purpose. I believe the free enterprise of corporate systems works in America,but I keep reminding myself that it was not ordained by God. In a democracy, any economic or social system can be legally altered by the people at the polls. So the people have a right to expect reasonable benefits, enlightened management, fair treatment and equitable distribution of wealth for the public good.

Quello, once he became Interim Chairman of the Commission, opened his regularly scheduled Monday morning meetings with his staff, a time when he was being privately briefed by top bureau personnel, to legal advisers and other commission staff members, so that aides to other Commissioners would know what information he had or was getting.He felt this involved all the FCC offices in the early information gathering and decision making stages of regulation. His supporters said it was something he had learned to be sensitive about, by his staff not being included in those meetings when he was a member all those years and not the Chairman.

Quello, in March 1993, came out in support of the escalating national efforts of activists to curb the amount of violence on television going so far as to say that he thought the FCC should have control over it the same way it did indecent programming, particularly in those hours when children were watching.

By Athena Smith-Dupre, Ph.D. candidate and Gerald V. Flannery, Ph.D.

ROBINSON, GLEN O.
1974-1976

Glen O. Robinson was born in Salt Lake city, Utah, on June 6, 1936, the son of Fern O. and Brigham H. Robinson. He attended Utah State University, then transferred to Harvard, where he earned a Bachelor of Arts Degree in 1958, graduating *magna cum laude*. He then entered law school at Stanford University where he earned his LL.B. degree with honors in 1961.

Robinson then took, and passed, the bar exam, and was admitted to the District of Columbia bar in 1961. He became an associate in the Washington law firm of Covington and Burling. That same year, he married D. Kay Costly.

After working with the firm for only one year, Robinson entered the United States Army and was honorably discharged two years later, in 1964 with the rank of First Lieutenant.

Robinson returned to the law firm of Covington and Burling and worked there until 1967. In September of that year, he became a professor of law at the University of Minnesota, joining the ranks of Supreme Court Justice Harry Blackmun and former Supreme Court Chief Justice Warren Burger, who had also held law professorships at the St. Paul, Minnesota campus. Robinson held that position until 1971.

On July 10, 1974, Robinson was appointed a Commissioner of the Federal Communications Commission in Washington, D.C. by President Richard Nixon. He was given the oath of office by fellow Minnesotan, Supreme Court Justice Harry Blackmun. Broadcasting magazine assessed the Commissioner:

> Professor Robinson--bright, young, and energetic--comes straight from the campus and that, without provocation, is enough to make him controversial. Anti cross- ownership and Fairness Doctrine comments attributed to him have aroused the interest of diverse groups

Robinson, in 1975 (in opposition to chairman Richard Wiley's desires), wanted the Fairness Doctrine repealed because he thought it might have a chilling affect on the licensee's disposition to present controversial material on television or radio.

While on the Commission, Glen O. Robinson said that he felt broadcasters should inform the public on media issues. Robinson also viewed the FCC Community Access Television (CATV) regulation as "excessive." Robinson expressed concern over FCC intrusion into programming. He was especially concerned over the need for FCC action regarding children's advertising.

The FCC's and Robinson's docket in 1975 was very busy. Broadcasting magazine reported his involvement and views on many issues: more VHF channels to aid Pay TV development, feeling Pay TV programming must serve the public, not groups. He wanted a broad newspaper-broadcast Community Access Television cross-ownership rule and backed the FCC's new cross-ownership rule in that area. He favored repeal of the Fairness Doctrine and backed the ruling making FCC meetings open to the public. He expressed concern over CATV Equal Employment Opportunity practices, and over the Commission's growing CATV regulation. Robinson served a total of two years, resigning from the Commission on July 30, 1976

During the first six months of 1976, Commissioner Robinson's final half-year on the Commission, he belittled the FCC's efforts in community ascertainment and talked about returning to teaching. However, this period proved to be Commissioner Robinson"s most active of his FCC career. During this period, he backed Chairman Richard Wiley on a five-year license period for radio and television broadcasters and he urged the FCC to apply the duopoly rule to Educational Television. Robinson was asked to head the Aspen Institute's communication regulation and policy area and was then named special advisor to the Aspen Institute. He favored a five-member FCC, instead of seven. In the July 26, 1976 issue of Broadcasting magazine, Robinson criticized the FCC on priorities fixing. He saw the FCC as protecting broadcasting against CATV but commended the FCC on cross-ownership actions. In July, Robinson became an associate of the Aspen Institute.

Robinson, once he left the agency, moved just outside Washington to McLean, Virginia, with his wife Kay. They had two children, Dean and Jennifer. Robinson then became a law professor at the University of Virginia.

Robinson gave a sort of exit interview to Broadcasting magazine in July 1976 as he prepared to go back to being a law professor, the sort of quiet contemplative life he liked. Asked if he thought the broadcast industry would be better off without the FCC he said he would have to think long and hard about that. He felt the agency was necessary in the technical areas it dealt with but sometimes , in other areas, it nit-picked things to death. He told <u>Broadcasting</u>, one of the things about an agency like the FCC is that it "...not only regulates its regulatees, it protects them. That's part of the regulatory life style. It's not a one-on-one, captive-of-the-industry situation. It's a great deal of identification; its evident in other agencies, too."

By Gerald V. Flannery, Ph.D. and James Nunez, M.S. candidate

WASHBURN, ABBOTT M.
1974 -1982

Abbott McConnell Washburn was a common sense moderate who exhibited a stabilizing effect on the Federal Communications Commission from 1974 to 1982. His eight years with FCC added needed communication knowledge and experience, for technology was on the cutting edge of an explosion that would change the communication field tremendously.

Washburn was born March 1, 1915 in Duluth, Minnesota. Spanning the time he graduated from Harvard until his presidential appointment to the FCC in 1974, Washburn's resume' resembles a who's who in the communication field. He was the Director of the Department of Public Services for General Mills in the 1940's. His Executive Chairmanship of the Crusade for Freedom (Radio Free Europe) in 1951 was followed by an assignment to Eisenhower's campaign staff in 1952. In 1953, he was named Deputy Director of the United States Information Agency where he served until 1961. That position was the springboard that brought Washburn recognition as a noted international communications expert. In the years that followed, he served as president of Washburn, Stringer and Associates, Inc., a public affairs counseling firm.

Washburn's deputy chairmanship of the U.S. delegation to the International Telecommunications Satellite (INTELSAT) Conference in 1969, led to the Chairmanship of the INTELSAT Negotiating Conference in 1970 with the rank of ambassador. He served as a special consultant to the director of the White House Office of Telecommunications Policy from 1970 until 1974.

President Richard Nixon appointed Washburn to the FCC on July 10, 1974 to fill a one year vacancy. Nixon's other two FCC appointments at that time, H. Rex Lee and Glenn O. Robinson, were presumed to be politically motivated. Washburn's credentials were unquestionable, and he was considered a worthy candidate. President Gerald Ford reappointed him and swore him in on October 9, 1975 for a full seven-year term.

Washburn's experience served the FCC well during those volatile years. One of his proudest accomplishments was his 1975 stand on reducing the minimum size of earth stations. Washburn made the dish reduction suggestion in a meeting at the home of Lionel Van Deerlin, a congressman and Chairman of the House Communications sub-committee. The Wiley Commission approved and acted on it in 1975. Washburn considered himself the "father of those dishes", a title he justly deserved. He wished he had a dollar for every satellite dish that was seen on the rural roads of America. The acceptance of his dish reduction idea opened a world of information for many. Television / Radio Age quotes Washburn as claiming the dish size reduction "is what led to the mushrooming of dishes, which led to pay-TV, HBO, Showtime and everything else. That was a very significant action by the FCC, and I'm glad I had a part in it. It really started an industry." Time proved him right, as cable television became a major media force in millions of homes worldwide.

Washburn's stands were of the common sense approach. He would look at each issue independently and judge it on it's merits. One such case involved Montgomery County Broadcasting, a company that decided to build a radio tower before the FCC granted it's permit. The company was trying to get operations started in the early fall, rather than the following spring. It took a big gamble that approval would be granted. The FCC staff recommended that if the tower was taken down, the FCC would then grant the permit. Washburn thought that this recommend-ation was "absolutely crazy", and "extremely bureaucratic". His views prevailed in a 4 to 3 vote by the FCC to allow the tower to stay and to grant the license to Montgomery County Broadcasting.

Washburn was in favor of the FCC licensing of low power television stations (LPT). He was an advocate for restrictions on multiple television station ownership. He also was for giving preferential treatment to non-commercial applicants. He stated that very few applications were from non-commercial applicants, and some preference must be given to them so that they could compete. Washburn was not convinced that a completely free and open market could sustain small limited-market programming.

Television / Radio Age, April 5, 1982, voiced Washburn's feelings concerning the Prime Time Access Rule. Washburn said "It just about doubled the number of independent producers...it has done what it set out to do. It's not true that it has just spawned game shows."

The issue that got away from Washburn was the AM stereo controversy. His lone dissension on the Commission's open marketplace decision on AM stereo was a disappointment to him.

Washburn noted that standards had always been successfully set for other media, why not AM stereo. He thought the open marketplace would doom it, noting that the competing systems were incompatible. AM stereo did not become the media force it could have been, perhaps, if certain restrictions and regulations would have been enforced. On this issue, as well as other important issues, Washburn exhibited great foresight in applying his views and opinions. His decisions, and explanations for those decisions, were based mainly on drawing from the vast knowledge and experience he possessed as well as his common sense.

A pet subject for Washburn was the issue of children's television programming. He felt that government must show some consideration of what he called an unique audience. The FCC's stand on self-regulation of children's programming was considered by him to be somewhat appropriate. Washburn did, however, oppose the recommendation of the Ferris Task Force that called for mandatory hours of children's programming. This call for strong regulation may have seemed too restrictive as it was stated. According to Washburn, the Republican FCC Chairmanship had completed a good policy statement that was approved unanimously by the Commission. It stressed programming with educational and entertainment value and provided incentives for good programming for children. Washburn was not impressed with the choice he then had on children's programming. He said that President Ronald Reagan's deregulation moves went too far. Washburn considered the reduction of FCC Commissioners from seven to five a mistake. He noted that the work load would be tremendous.

Broadcasting magazine called Washburn "an old line conservative, but no sure vote in any Chairman's pocket . . . he has never been anyone's automatic vote . . . won't be Fowler's either." In 1981, with Reagan's presidency striving for deregulation with the appointment of Mark Fowler as FCC Chairman, Washburn was not considered for renomination. His term expired in 1982.

Washburn, after he left the FCC, began serving on the boards of several companies. Much of his time was spent doing philanthropic work with the Eisenhower Institute. He also worked on the President's Task Force on International Broadcasting which made recommendations on U.S. Government policies with regard to international broadcasting operations. This task force issued guidelines on handling international broadcasts in the wake of the Cold War.

By Richard Tanory, M.S. and Gerald V. Flannery, Ph.D.

WHITE, MARGITA E.
1976-1979

      The road to success was a long and interesting journey for Margita White. Born in Linkoping, Sweden, on June 27,1937, she became a naturalized citizen in 1955. She attended the University of the Redlands, in San Bernardino, California, where she was homecoming queen and a *magna cum laude* graduate with a B.A. in government. Her next stop was Rutgers University, New Brunswick, N.J., where she received an M.A. in political science in 1960.

      Graduation from Rutgers marked the beginning of two careers. She became assistant to the press secretary for the Richard Nixon presidential campaign (1960) as well as becoming the wife of lawyer Stuart Crawford White in 1961. From that point her career became quite a balancing act.

      She functioned as an assistant in the Whitaker & Baxter advertising agency from 1961 to 1962. The following year, White returned to the political scene as the minority news secretary in the Hawaii House of Representatives. She moved on to become a research aide to Senator Barry Goldwater and research associate to the Republican National Committee from 1963 to 1964. The following two years were spent as a research assistant and writer for the Free Society Association. For the duration of 1967, White acted as assistant to syndicated columnist Raymond Moley.

      White worked for the Nixon administration from 1969 to 1973 where she was an aide to Herbert G. Klein, the first director of the Office of Communications. During the Watergate years, she held a position as assistant director for the public information at the U.S. Information Agency. The following Year, 1975, she assumed the position of assistant press secretary to President Gerald Ford. Her responsibilities in that job were to coordinate presidential press conferences, monitor the public relations activities of the executive branch departments, and to distribute information to Washington correspondents and the media.

      After serving for one year as the director of the White House Office of Communication, White was nominated on July 22, 1976, to the Federal Communications Commission to succeed Glen O. Robinson. However, her nomination met opposition from the Senate during the August 25th confirmation hearing. Several committee members claimed that a conflict of interest existed over White's husband's legal activity. Mr. White, a lawyer, represented clients whose cases frequently went before the FCC. Mr. White subsequently left his law firm on March 14, 1977 paving the way for Mrs. White's Senate approval on September 9th of that year.

White spent 29 months on the FCC until her resignation on February 28,1979.   During her tenure she was noted for being straightforward. She was the only vote for a 7th trans-atlantic cable (TAT-7) in 1978; the Commission accepted the plan the following year. She was the first Commissioner to press for a flexible approach to the 1979 World Administrative Radio Conference in order to allow the FCC to share UHF channels with land mobile and fixed radio services.

She openly criticized the stereotyping and sexual exploitation of women on television and she urged parents to supervise the television viewing of children.

In an address in October 1978, White advised broadcasters to look to Congress for de-regulatory relief. She thought the greatest threat to an independent broadcast media came from the Commission itself. She openly opposed the plan to rewrite the Communication Act provision calling for a spectrum use tax whose proceeds would be used to support public broadcasting. In her last speech as member of the FCC, White said she felt that the system was stacked against Commissioners because of inadequate time and resources to thoroughly explore alternatives to staff recommendations. In addition, she believed that the Chairman had too much influence over policy decisions. White did not have the academic or legal background that other Commissioners did when she was appointed but she was regarded as intelligent and effective; she did however have media experience mainly as a writer and through her work in political and governmental public relations.

The journey did not end with her resignation from the FCC. White moved on to Caucus Incorporated as a coordinator, then to ITT, Taft Broadcasting, Rayoneir Forest Resources, and Golden Jubilee Communication Telecommunications.   In 1982 she was the U.S. delegate at the International Telecommunications Plenipotentiary Conference. For a time, White was a member of the Washington Woman's Forum and member of the George Foster Peabody Advertising Board (1979-1986). Then followed a post with the U.S. Information Agency as a member of the advisory committee. Her road to success was also paved with several special honors. In 1974, she received the distinguished service award from the University of Redlands Alumni Association. The following year she was honored as a member of Executive Women In Government, as well as receiving the Superior Honor Award from the U.S.I.A.

Margita White had two children, Suzanne Margareta and Stuart Crawford. She and her husband made their home in McClean, Virginia.

By Gerald V. Flannery, Ph.D. and David A. Male, M.S.

FERRIS, CHARLES D.
1977-1981

Charles Daniel Ferris was born on April 19,1933 in Dorchester, Massachusetts. He graduated from Boston College with a major in physics in 1954. In 1955, he joined the Navy where he worked as a Chief Engineer. He stayed there for five years and in 1961, he entered Boston College Law School and enrolled in the Honors program.

In 1962, he became a trial attorney for the Civil Division of the United States Department of Justice. Ferris then moved to Congress where he was a Congressional aide from 1964 until 1977. He served first as Assistant General counsel to the Democratic Policy Committee; then moved to be Chief Counselor to the Senate majority leader, and later became Senator Mike Mansfield's Chief of Staff. He held offices in Congress for nearly 14 years and his power became so great that he was regularly called the 101st Senator.

Ferris worked long hours but he never forgot to touch all the bases. Congressional reporters remember him regularly coming up to the press gallery for a game of gin rummy or hearts.

Before he entered the Federal Administrative system, Ferris had no experience in that area. However, his reputation in the offices of Congress for being fast and decisive helped him greatly all throughout his Chairmanship at the Federal Communications Commission. One of the problems he saw when President Jimmy Carter named him Chairman of the FCC on October 17, 1977, was that the Commission's employees were rigidly following an 8:00 a. m. to 4:30 p.m. working day. As the new Chairman, Ferris felt that early closing hours represented a hardship to the public, especially on the West Coast, where the time difference made it only 1:30 p.m. at closing time. So as one of his first official acts, Ferris decreed that the working day would run from 9:30 a.m.-5:30 p.m. There were, of course, loud protests, but the new Chairman was a clever negotiator. He agreed to have the change delayed 30 months but he eventually got his way. The Commission's working day by the time he left the post (1981) was running 8:00 a.m. to 5:30 p.m.

After his first clash with his staff members about the office hours, Ferris cautiously eased into the role of Chairman. He did not give his first public speech for six months, but when he finally did, it became obvious that his FCC was on a different road. Under Ferris, the Commission would first have to see if its intrusion as a regulatory body in broadcasting was really called for.

Ferris once said the FCC must take a hard look at whether the market forces could more effectively serve the public interest before it decided to regulate. That policy led the Commission to move closer to removing the restrictions that required all radio stations to limit their commercials and to devote a certain amount of time to news and public affairs. His theory was that stations would be able to provide diverse, innovative broadcasting without being stifled by rigid rules. On the other hand, Ferris was not an advocate of life-time ownership of station licenses.

On February 1, 1978, Ferris attended his first Congressional hearing before the House Appropriations Subcommittee. It dealt with the Commission's proposed budget of $67,035,000 for the fiscal year 1979. It lasted only two hours, but many issues about the Commission's future under Chairman Ferris were discussed. He indicated that he intended to revise the children's television proceedings to determine whether broadcasters were meeting the regulations set out for them in the FCC's policy statement on children's programming. Broadcasting magazine quoted Ferris, talking about television programming for children, as saying

> As a parent, I do have a responsibility, but as a Chairman of the FCC, I don't know how I could enforce standards without violating the Constitution. And as much as I might have strong feelings of revulsion toward certain programs, I have strong feelings about the First Amendment.

During his first year in office, Ferris conducted an investigation into allegations that the television networks exercised undue influence in the industry. He appointed two academicians in the capacity of co-directors of that investigation. On March 1978, Ferris approved a new regulation on licensing renewals of television stations. According to the new rule, when hearing a case about a license renewal, the Commission had to base its judgement on programming grounds alone.

Ferris once said he had only a few goals for the FCC. They were: 1) deregulate where the marketplace will work effectively; 2) increase the focus of the Commission on common carrier matters; and 3) involve more non-lawyers, economists and technical people in the decision making process.

After serving his term in office, Chairman Ferris went back to practicing law on April 14, 1981, with the firm of Mintz, Levin, Cohn, Glovsky and Popeo.

By Marisol Ochoa Konczal, M.S. and Gerald V. Flannery, Ph.D.

BROWN, TYRONE
1977-1981

Tyrone Brown was born in the Tidewater area of Virginia on November 5, 1942. His parents lived on sharecropper farms before moving north to New Jersey during World War II in search of a better life. Brown was one of seven children, and money was scarce, but with both parents working the family was able to make ends meet.

Brown proved early in his education, that his goals were set high. At East Orange High School, he was a top student, an athlete, and the first black president of the student council. In 1964, he received his B.A. with honors from Hamilton College in Clinton, New Jersey; and while attending Cornell University Law School, he was the editor of Cornell Law Review and received the Frazer Prize for leadership and academic achievement. In 1967, he received his LL.B. with distinction.

Creating a professional career that could live up to his educational background would seem a difficult goal, but Tyrone Brown was up to the task. He set the pace early when, in 1967, Brown served as a law clerk to Chief Justice Earl Warren. From there his resume grew to include positions such as Associate Attorney with a Washington law firm from 1968 to 1970, Special Investigator in Jackson, Mississippi, for the President's Commission on campus unrest in 1970, assistant to Senator Edmund S. Muskie from 1970 to 1971, and Director and Vice-President for legal affairs of Post-Newsweek Stations, Inc., from 1971 to 1974.

This last position with Post-Newsweek, was an important one because it moved him well up the career ladder and gave him important experience and knowledge in the field of broadcasting. After leaving Post-Newsweek, he joined Caplin and Drysdale, a tax law firm in Washington, D.C. It was at this time, September 1977, that President Jimmy Carter said he was prepared to nominate Brown to a vacant seat on the Federal Communications Commission.

At first, Brown refused the appointment because it was only for a two year term. Carter had just named Charles D. Ferris to a seven year term and, according to The New York Times, the Congressional Black Caucus had let it be known that they wanted Brown to get that assignment. A few days later, however, Brown changed his mind and decided to complete the term that had been left open by the resignation of Benjamin Hooks. Brown took the position after key politicians urged him to do so and President Carter gave him reason to believe he would be reappointed to a full term.

There were few questions at Brown's Senate Commerce Committee hearing, in fact, according to Broadcasting magazine, the entire affair only took about 35 minutes. On November 15, Brown was sworn in as a member of the FCC.

In October 1977, Brown was quoted in a Broadcasting magazine article regarding a set of principles he said would guide him as a member of the FCC:

> I intend to be equally accessible to all interests, industry and non-industry, and I'll keep an open log.   I intend to take every precaution to guard against a conflict of interest. I believe competition should be encouraged whenever it's a viable alternative. I strongly oppose   government regulation   or censorship of program content. I don't believe in regulation for the sake of regulation... Minority groups are entitled to full participation in broadcasting, in terms of employment and ownership.

With these clearly stated goals, Brown began his term as an FCC Commissioner.

About a year later, in November 1978, Broadcasting ran a profile of Brown. It was a chance to see if he was living up to his background-- and he was. Attorneys practicing before the agency described him as one of the best FCC Commissioners of all time. Brown did, in the article, set two new goals for himself. The first was to try and move away from the trend of concentrated media ownership and the second was to work for greater involvement of minorities in the communications industry.

Broadcasting magazine, in its latest chronology of important events, listed a number of things the FCC dealt with during that time. Some of them were: Warner began its historic two-way interactive television system (QUBE) in Columbus, Ohio; the U.S. Supreme Court agreed to review the famous "seven dirty words" case; President Jimmy Carter signed into existence the National Television Information Administration; the agency established a policy aimed at promoting minority ownership in broadcasting; the U.S.Supreme court ruled that the FCC did not have the power to force cable systems to make access channels available; it removed the limitation about cable systems importing distant signals; the agency began what would result in sweeping regulation reforms for radio; Ted Turner began CNN, the 24-hour news service; Public Broadcasting underwent a major reorganization; the FCC ruled on children's programming; and FCC approved low power television.

In 1981, Brown became one of the first politicians to be affected by the changing presidency. With Ronald Reagan, a Republican, coming into office, Brown submitted his resignation effective January 31, 1981. He went on to join the Washington law firm of Steptoe and Johnson, with interests in many areas including communications.

By Gerald V. Flannery, Ph.D. and Richard E. Robinson, M.S.

FOGARTY, JOSEPH R.
1976-1983

Joseph R. Fogarty jumped into the Federal Communications Commission with both feet when he was appointed in September of 1976. Controversy over fellow nominee Margita White's conflict of interest earned him the seven year term rather than his original two year appointment. Fortunately for the FCC, Fogarty came to the post well prepared after serving for the Senate Communications Subcommittee the previous year and its parent Commerce Committee for eleven years.

In questioning before the Senate Commerce Committee prior to his appointment, Fogarty was outspoken. He felt that FCC meetings should be open to the public; that contributions by citizens to the FCC proceedings should be reimbursed; that the FCC should encourage cable television growth; require licensees to cooperate with citizens' groups, and provide broader access to the broadcast media.

In a 1982 interview, Fogarty's focus may have shifted, but his liberal determination held fast. Technology changed over the course of his six years on the FCC and his concern centered on authorization of broadcast satellites, and in association with his area of expertise, common carrier regulation. Early in his career with the FCC, Fogarty saw the challenge of computer communications, data processing, and fiber optics as the future of communications. Thus, Fogarty felt that it was fair to reevaluate the multiple ownership rules facing broadcasters.

When the rules limiting ownership of various media were initiated, the diversity of technology did not exist. Fogarty felt that without the ability to get into the game, the common broadcaster stood no chance at fair play. FCC regulation prevented the industry from competing in the new markets.

Fogarty remained concerned about the public interest, but the emphasis changed to AT&T. His opinion was that the FCC retained jurisdiction over the service which through power granted by the Communications Act, put the public interest first. He also felt that broadcasters deserved greater scrutiny at license renewal time to preserve the public service responsibilities of the medium.

In 1983, Joseph Fogarty left the Commission to head the Telecommunications Department of the New York law firm of Weil, Gotshal and Manges. It was his first time back in private life since his government service began in 1963.

Born in 1931 in Newport, Rhode Island, Fogarty attended Holy Cross College where he received an A.B. in 1953. From there he joined the Navy from 1953 to 1959 and attended night school to obtain his law degree from Boston College in 1959. Showing sheer determination, he traveled 180 miles round trip three or four nights a week to complete that degree.

From 1959 to 1964 he practiced law with firms in Massachusetts and Rhode Island. In 1963, Senator John Pastore, a Democrat from Rhode Island, hired Fogarty to assist him in various Commerce Committee matters. Although he was practicing law, Fogarty had also participated in local Democratic politics and was pleased to accept the Senator's offer.

For eleven years, Fogarty worked as Staff Counsel for the Commerce Committee, until 1975. He was responsible for transportation legislation and also east-west trade policies. In this regard he learned the Russian language.

In 1975, he moved on to the Communications Subcommittee where he advised Senators on telecommunication policy matters. He also assisted n preparing committee reports for communication legislation. In this position, Fogarty was able to make contacts with members of the FCC, Office of Telecommunications Policy Board, and common carrier industry representatives.

Broadcasting magazine, in its latest chronology of important events, listed a number of things the FCC dealt with during that time. Some of them were: the FCC voted to retain its equal time law; President Jimmy Carter signed into law the National Telecommunications and Information Administration; the U.S. Supreme Court ruled that the FCC did not have the power to force cable systems to make access channels available; it removed the limitation about cable systems importing distant signals; the agency began what would result in sweeping regulation reforms for radio; Ted Turner began CNN, the 24-hour news service; Public Broad-casting underwent a major reorganization; the FCC ruled on children's programming; the FCC approved low power television (LPT); the FCC decided the open marketplace was the best regulator; the FCC used a lottery to choose among the 12,000 applicants for low-power licenses; it allowed FM stations to use their sideband space to market special services; and it opened up the FM spectrum creating three new classes of stations.

Fogarty married Joan Baxter and they had six children. Over his eight year period in the communications field, first with the Senate Subcommittee and then the FCC, he became an expert on telecommunications. Most of his knowledge came from self education and a determination to understand the concepts. This attitude served him well during his years with the FCC.

By Gerald V. Flannery, Ph.D. and Peggy Voorhies. M.S candidate.

JONES, ANNE P.
1979-1983

When referring to specifics, Anne Patricia Jones   public pronouncements did not reveal all aspects of her thinking. As indicated in Telephony, to some extent she kept her own counsel, reached her own decisions and marched to the sound of "her own drummer."

Jones was born in Somerville, Massachusetts on February 9, 1935, the daughter of William C. and Helen Donnelly Jones. She married William Sprague on June 20, 1981. Originally, Jones planned to pursue a secretarial career in Boston. When registering for one night class at Boston College, she was persuaded by Father Paul Ryan, then dean, to take a full load of classes. After attending night, weekend, and summer classes, she received a Bachelor of Science (*magna cum laude*) five years later. Three years later, she graduated from Boston College Law School (*cum laude*), where she was a member of the esteemed legal fraternity, Order of Coif.

From   July 1961 to April 1968, Jones practiced general corporate law with the Boston law firm of Ropes & Gray. She then put her legal mind to work with the Securities and Exchange Commission from April 1968 to January 1978. During this ten year span, Jones consistently was promoted and advanced. While at the SEC, she was an Attorney Adviser (April 1968-April 1969) and Special Counsel (April 1969-July 1970) in the Division of Corporate Regulation; Legal Assistant to Commissioner James J. Needham (July 1970-July 1972); Associate Director (August 1972-January 1976);   and Director of the Investment Management (January 1976-January 1978).

In January 1978, Jones was appointed General Counsel for the Federal Home Loan Board . Originally, she planned to stay in Washington only two years, but her steady move up the government ladder caused her to extend her stay. One year after being named FHLB General Counsel, she was nominated by President Jimmy Carter to succeed Margita White as an Federal Communications Commissioner.

Jones glided through her confirmation hearing on February 23,

1979. It was apparent that in the fifteen minutes she was before the Senate Commerce Committee, Jones had researched the 22 questions issued prior to the beginning of the hearing. At this time, Jones did not clearly indicate what kind of Commissioner she would be, but she did express her hope to see television offering "intellectual stimulation" and radio offering "greater diversity."

However, Jones' hearing was not without controversy. Making it clear their concerns went beyond the nominee, two Hispanic-American women strongly opposed the nomination. Sharleen Maldonado and Nelda Ojeda Wyland felt the vacant slot belonged to them. Their feelings stemmed from an inability to persuade the government to address their problems and those of the "Hispanic telecommunications industry." Down the road, however, two Hispanic members would later join the Commission, Henry Rivera and Patricia Diaz Dennis. On March 21, 1979, the nomination of Anne Patricia Jones was confirmed by the Senate. Twelve days later, she was sworn in as an FCC member. Jones broke the usual tradition of taking the oath in Washington; she was sworn in where she claimed it all began -- Boston College.

Chairman Charles D. Ferris and Joseph Fogarty, both FCC members, were present; as well as U.S. District Judge David Nelson who administered the oath. Chairman Ferris saw the event at Boston College as "one of the most delightful occasions" in he had participated in. This statement disguised the rumor that Jones, although Republican, had been chosen to assure Ferris a 4-3 Democratic majority vote on the commission. For it was no secret that they has been classmates at Boston College Law School and that he had suggested her to the Carter White House.

Jones quickly demonstrated she was not in the pocket of Charles Ferris. She did not lead on many issues at the Commission; for her style, as Broadcasting magazine quoted fellow commissioner Mimi Dawson, as saying, was to "serve as the conscience of the Commission." Jones connected with the Commission at a time when the pace of regulatory reform was accelerating. She was a strong believer in reviewing older FCC regulations and getting rid of those that didn't make sense then.She felt that sometimes regulations hampered new technology making it unprofitable for developers to enter the market. For example, she said, the standards put in place when color television was being introduced actually hampered its development because the FCC chose the wrong standard.

Quickly, Jones began a mission to support high standards for children's television, but without ordering special programming. A staff report on children's programming guidelines, adopted by the

Commission in 1974, found that "television had failed to respond adequately to the needs of its child audience." Jones worried that the FCC might go too far in encouraging better programming by adopting mandatory laws. She said she would have to be persuaded that the Commission should mandate programming.

Jones classified herself in the deregulation camp. Television / Radio Age quoted her as saying,"I've been with government long enough to know that government regulation isn't the answer, doesn't provide the answers to all the world's problems. Actually, I think maybe the less, the better." She strongly believed in examining past FCC regulations and discarding those that made no sense. She believed in the marketplace theory, which states that the marketplace is he best regulator of programming, that no protectionism should be permitted in the government.

In an article in Television / Radio Age, Jones went on to say: "That's why I'm in favor, after studying all the consequences, of doing away with our cross-ownership rules in broadcasting. And things like that. We should allow the experts from all fields to come in." In the same article, she went on to say, "Let there be competition. But it (the Commission) seems also to be saying: Let there be competition among the people who are already in." Jones wanted stronger steps to allow entry, to permit more players.

Almost as soon as she took office, Jones won a reputation for public candor. On the fifteenth anniversary of the day she began working for the government as an attorney-adviser for the Securities and Exchange Commission, and after just over four years as a Commissioner at the FCC, Jones announced her intention to resign.

In Broadcasting magazine, Howard Monderer, vice president and Washington attorney for NBC, called her a "terrific" civil servant he was sorry to see go. "She brought a sense of inquiry to everything she did," he said. Also in Broadcasting magazine, Eugene Cowen, ABC Washington vice president, said he had enormous respect for her, even though her votes did not always favor the broadcasting line. "She is really an independent mind," he said. "One of the finest Commissioners ever."

It is with great regret and guilt that Jones resigned, particularly concerning the advancement of equal opportunity, regulating AT&T, and guiding telecommunication policy-making. Jones was most proud of a 72-page statement outlining an alternative to the FCC's access charge decision for the telephone industry.

Broadcasting magazine, in its latest chronology of important events, listed a number of things the FCC dealt with during that time. Some of them were: the U.S.Supreme court ruled that the FCC did not have the power to force cable systems to make access channels available; it removed the limitation about cable systems importing distant signals; the agency began what would result in sweeping regulation reforms for radio; Ted Turner began CNN, the 24-hour news service; Public Broadcasting underwent a major reorganization; the FCC ruled on children's programming; the FCC approved low power television (LPT); the FCC decided the open marketplace was the best regulator; the FCC used a lottery to choose among the 12,000 applicants for low-power licenses; it allowed FM stations to use their sideband space to market special services; and it opened up the FM spectrum creating three new classes of stations.

She planned to revert to the private sector, remaining within the law field, but focussing more on communication. Broadcasting magazine said, at age 48 Jones felt :

Change is good. I think everyone should change occasionally because a lot of fear goes with change, she said. But I think it sharpens you. . . It keeps you as sharp as you can be. You can let up. When you get used to things, you can take them for granted.

Jones' drive to confront issues head on never left her. Even on her final meeting day, she arose at 5:30 a.m. to complete her review of the agenda. A term that would have officially ended June 30, 1985 came to a halt on May 31, 1983.

By Kelli M. Vallot, M.S.and Gerald V. Flannery, Ph.D.

SHARP, STEPHEN A.
1982-1983

Stephen Alan Sharp had one of the briefest tenures on the Federal Communications Commission. After a nine-month stay, Sharp left on June 30, 1983, the victim of a political maneuver that cut the panel from seven members to five, leaving Sharp and fellow Commissioner Joseph R. Fogarty out in the cold.

Sharp wasn't embittered by that experience. Indeed Sharp, who was 36, seemed relieved for financial reasons to be moving along. He told _Broadcasting_ magazine: "I regret not being independently wealthy so I could devote my entire life to being a dollar-a-year- man... Public service is interesting, but its not that interesting."

The magazine succinctly explained how Sharp, a Republican, was ousted saying [his removal was] the result of a political tug of war between the Republican controlled Senate Commerce Committee and the Reagan White House. The committee had refused to hold a confirmation hearing for Sharp for several months in deference to Senator Ted Stevens (R-Alaska) who was upset when the White House passed over his own candidate for the Commission. The committee subsequently short-sheeted Sharp, cutting his term to nine months in legislation   trimming the Commission from seven members to five, a move some observers said was aimed as much at saving the committee face as it was at paring the   federal budget.

Though Sharp spent only nine months as FCC Commissioner, he had been working with the agency for about 10 years as an attorney prior to his Commission appointment, which came when he was 35 years old.

Sharp was born in Columbus, Ohio, on June 10, 1947, the son of William George Sharp, Jr. and Barbara Martin Baugham Sharp. He was an Episcopalian.   After graduation from high school, he attended Washington & Lee University in Virginia, graduating in 1969. He earned his law degree in 1973 from the University of Virginia and was a member of Phi Sigma Alpha and Delta Theta Phi.

While attending school, Sharp, worked as a reporter and an announcer for radio stations WWST and WWST-FM in Wooster from

1965 to 1967, as an announcer for radio station WREL in Lexington, Virginia from 1967 to 1969, and as a reporter for WAVY-TV in Portsmouth, Virginia in 1968. After graduating from Washington & Lee University, he was the director of communications for the Dunn for Governor Campaign in Nashville, Tennessee in 1970, subsequently becoming news secretary and special assistant to Governor-Elect Dunn.

His affiliation with the FCC began in 1972 when, as a law student, he became a legal clerk and then later a staff attorney, for the Commission's general counsel, from 1974 to 1976. During this period, he also worked as counsel for Impeachment Inquiry, the U.S. House of Representative's Committee on the Judiciary. From 1976 to 1978, he was legal assistant to FCC Commissioner Margita White.

IN 1978, he became an associate of the Washington law firm of Schnader, Harrison, Segal & Lewis, with which he was affiliated until 1981. In that year, he returned to the FCC as its general counsel, a job he held for a year until being sworn in as a Commission member on Oct. 5, 1982. Up until this time, he also had been a adjunct professor of law at George Mason University since 1977, as well as guest lecturer at the University of Virginia School of Law in 1979 and 1980. Industry observers often called Sharp the "eighth Commissioner" because his legal counsel successfully kept the FCC out of court on a number of issues.

Sharp's military service began in 1970 when he joined the Army Reserves as a Second Lieutenant and gradually rose in the ranks to become a Major in the reserve's Judge Advocate General Corps; and had been decorated with the Army Commendation medal.

As he was leaving the Commission in 1983, Sharp shared with Broadcasting magazine his feelings about the government's role in the regulation of radio and television. Sharp said he is a believer in regulation by the marketplace.

> It's not fair -- if we're going to be throwing in more TV stations and more radio stations and to let all this competition for broadcasters in - -that these broadcasters shouldn't be given some measure of freedom to deal with competitive changes..... So the deregulation side of the equation was brought to bear.
>
> The marketplace will be as open as possible. That's how I see it. It's starting now. It will continue as long as Mark Fowler is chairman, and then continue with his successor, no matter what party. That's the drift of history. Technology is driving this; it's not politics. The politics recognizes the technological change and adapts to it.

Sharp said one of his leading accomplishments while with the Commission was the Track One legislative package, which was put together while he was general counsel. Much of the proposal was voted into law, and "It's the first bill in 20 years initiated by the Commission that passed congress," Sharp said. He also listed the FCC's decision to stop licensing of citizen's band radios as a high point of his tenure.

Broadcasting magazine, in its latest chronology of important events, lists a number of things the FCC dealt with during Sharp's tenure. Some of them were:the FCC authorized an experiment using a direct broadcast satellite; two radio stations, one in San Antonio and the other in Pittsburgh, began broadcasting in AM Stereo; the FCC decided the only fair way to award new licenses to over 12,000 competing applicants was to use a lottery system; the FCC authorized FM stations to use their subcarrier channels to provide other services; and the agency opened up additional FM spectrum space, allowing three new classes of stations.

What doesn't Sharp miss about the FCC?

...the non-money and the bureaucratization... Sometimes things just don't happen. If you want to get something done, it's like pushing string. On the other hand, he'll miss the people at the Commission. It's a good staff; these people are good people... I'll be working with them in a sense from the outside in the future, but it's not the same.

After leaving the FCC, Sharp continued to practice administrative and regulatory law in the Washington, D.C. area.

The Sharp family lived in Alexandria, Virginia, and included Sharp's wife, the former Lynn Cawley, and their daughter Sarah.

By Michael A. Konczal, M.S. and Gerald V. Flannery, Ph.D.

RIVERA, HENRY
1981-1985

Henry Rivera was 35 years of age when he was appointed to the Federal Communications Commission by President Ronald Reagan. He was sworn in on August 10, 1981 and became the first Hispanic American to serve on the Commission. Rivera, a democrat, had no background in communications, but was described as a person who learned quickly.

Rivera, who was born in Albuquerque on September 25, 1946, had a BA in economics from the University of New Mexico and a law degree from the same institution. He worked for the law firm of Sutin, Thayer and Browne, and while in Albuquerque served as vice-president, president-elect, and as a member of the board of directors of the Albuquerque Bar Association. He was a member of the New Mexico Supreme Court Committee on rules of civil procedure and trustee of the New Mexico Law Foundation. He was President of the young lawyers division of the New Mexico State Bar Association, and a state delegate to the American Bar Association young lawyers division. He was editor of the Natural Resources Journal published by the law school, was Chief Justice of the Graduate Students Association court, and vice president of the Student Bar Association. For his service in Vietnam, he was awarded Army commendation medals and the Bronze Star.

Rivera came to the FCC as Congress passed legislation extending radio licenses to seven years and television licenses to five years. Congress also authorized giving the Corporation for Public Broadcasting $130,000,000 annually for the years 1984, 1985 and 1986.The day he was sworn in the FCC, in a special waiver, allowed CBS to own small cable systems, an experiment that would die a short time later.

Rivera, in an interview with Television / Radio Age in April 1982, said he was pessimistic about what the FCC would do in three areas: 1) multiple ownership; 2) the way the FCC handled equal employment opportunity requirements for broadcasters; and 3) its recent decision on awarding some 4,000 low power television stations. He was dedicated to creating more minority ownership opportunities in the broadcasting industry, but felt that the FCC's recent decisions in that area, particularly the NAACP's request that the Commission break up AM-FM combinations to provide more ownership opportunities, indicated it would not make changes in the multiple ownership rules. The ruling in the NAACP instance, a four-three vote, seemed, to Rivera, a prediction of how the FCC would vote on similar issues in the future.

Just ten days after Rivera joined the FCC, and before he had a chance to affect the decision, the agency, under the leadership of Chairman Mark Fowler, sent a major policy change statement to Congress, namely to repeal the reasonable access provision, the Equal Time rule and the fairness provisions of the Communications Act. Two weeks after he joined the FCC, Westinghouse purchased the Teleprompter Corporation for $646,000,000 making it the largest merger in the history of cable television. Rivera worried about similar situations in the broadcast industry where, without multiple ownership rules, minorities would be unable to raise the capital necessary to compete with conglomerates or established broadcast chains.

Rivera got heavily involved in the Advisory Commission for Minorities in Communication and spearheaded the fight with the White House about relaxing the EEO rules on hiring. The Chairman of the Commission, Mark Fowler, led a drive to deregulate the broadcast industry and to rely on a "marketplace theory" of regulation. That theory postulates that the marketplace is the best regulator of telecommunications services, and that the competition among stations and owners, will lead stations to better serve the public and to provide a range of programming that offers something for every type of audience. Rivera often supported Fowler in his actions.

Broadcasting magazine, in its latest chronology of important events, listed a number of things the Commission dealt with during Rivera's tenure. Some of them were: NBC's move to presenting an hour-long evening network newscast, something eventually rejected by the affiliates; the Gannett Corporation began publishing a national newspaper called USA Today; Turner Broadcasting began its news headline service CNN2; the FCC authorized an experiment using a direct broadcast satellite; two radio stations, one in San Antonio and the other in Pittsburgh, began broadcasting in AM Stereo; the FCC decided the only fair way to award new licenses to over 12,000 competing applicants was to use a lottery system; the FCC authorized FM stations to use their subcarrier channels to provide other services; the agency opened up additional FM spectrum space, allowing three new classes of stations; in July 1983 the FCC began looking for ways to deregulate television, and within a year, had begun to;   the FCC reiterated its belief that broadcasters had a responsibility to provide programming for children but left it up to the stations to decide how to do that; Congress passed a Cable Telecommunications Act; station ownership rules were set at 12-12-12. Rivera stepped down on September 15, 1985.

By Gerald V. Flannery, Ph.D.

FOWLER, MARK S.
1981-1987

Mark Stapleton Fowler was named to the Federal Communications Commission by President Ronald Reagan and confirmed by the Senate on May 14, 1981. He was born in Toronto, Ontario, Canada, on October 6, 1941 with dual citizenship, American and Canadian, son of an American father and a Canadian mother; later he elected American citizenship. He married Jane Yusko on August 18, 1963 and they had two children, Mark, Junior and Claire.

He brought to the job ten years of broadcasting experience beginning as a part-time announcer at radio stations in Florida during high school, such as WAVR-AM in Winter Park, and WHOO-AM-FM in Orlando; he continued that as a student at the University of Florida working for WDVH (AM) in Gainesville, where he graduated in 1966. After graduation he worked full-time as an announcer for WKEE-AM-FM in Huntington, West Virginia (1963-64); as an announcer and sales representative for WMEG (now WMEL) in Melbourne, Florida (1964-65); then as an announcer, sales representative, program director and production manager at WDVH while he went to law school at the University of Florida in Gainesville, from 1965-1970. After graduation he joined the firm of Smith and Pepper in Washington, D.C.

Fowler, before joining the FCC, was a senior partner in the communications law firm of Fowler and Meyers in Washington, D.C. He was the communications counsel to the Reagan for President Committee in 1979 and 1980, the Reagan-Bush committee during the 1980 campaign; also serving the Reagan campaign committee in 1975-76. He headed the FCC transition team after the 1980 election and was co-director of the Legal and Administrative Agencies Transition Group, Office of the President-Elect. He came to the agency prepared to deregulate and to encourage change in the broadcast industry.

Fowler, a Republican, was a strong believer in the power of the marketplace to regulate broadcasting, feeling it was more likely to provide for the needs of its listeners and viewers than the government could. One of the things he often said was "Let the marketplace decide." It was surprising then to discover that Fowler considered himself a liberal, someone who had worked for Bobby Kennedy in 1968, but did not vote for him, and who had voted for Lyndon Johnson for president, when his opponent was conservative Barry Goldwater. He did work for Reagan as part of his famous "kitchen cabinet," but felt his choices were driven by the issue and the person, not a hide-bound political philosophy. His work on the Commission bore that out as he tried to open doors to new technologies and allow them to compete for existing services while encouraging new technologies and new services. His brand of conservatism did not mean protecting the "status quo." He strongly supported, despite surprising opposition from broadcasters, doing away with the Prime Time Access Rule, which prevented the networks from controlling that period, just before the three-hour prime time block of programming began each evening.

Critics challenged Fowler's marketplace theory by saying that it presupposed competition but that in many markets there really was not competition, that the market had settled in to a certain pattern of " there's enough for everybody" and that often stations did not go head to head with each other or with newspapers or cable. Fowler felt that was a temporary condition that would soon change by the introduction of other technologies and services. The Commission had already endorsed the Ferris proposal to create a host of new low-power television stations (LPT), and the agency authorized AM Stereo, teletext, and Direct Broadcast Satellite services.

Television / Radio Age, in April 1982, said this about Fowler:

> No FCC Chairman in recent history has been so committed to a particular philosophy of regulation -- as Mark Fowler. His tireless reiteration of the doctrine that the marketplace can solve most -- if not all -- problems and his unswerving support for a pristine interpretation of the First Amendment make him an unusually predictable...

Fowler felt that broadcasting had been treated in a way that no other medium (print or film) had when it came to freedom of expression, or the First Amendment, because for fifty years it had been so closely regulated by the FCC, using the "scarcity" philosophy to justify its actions. He felt this was tantamount to saying that because broadcasting was the most effective type of speech it should be the most heavily controlled. To those who said he was merely a spokesperson for the broadcast owners, he said don't just listen to me, read the First Amendment.

Fowler said the Commission should have three objectives. First, to move to an unregulated, highly competitive stance which would encourage marketplace factors to serve as regulators; two, to eliminate antiquated, unnecessary rules and regulations, paperwork, forms and reporting rules, that were costly to follow and largely ineffective; and three, to expedite the internal and external functions of the FCC so that decisions got made quickly and fairly.

By Gerald V. Flannery, Ph.D.

DAWSON, MARY ANN
1981-1987

Mary Ann Weyforth Dawson was born in St. Louis, Missouri on August 31, 1944. She was the daughter of Francis Griffin and Jeanne Gething Weyforth. Mary Ann (Mimi) Dawson was reared in St. Louis and attended Catholic schools there. In 1962 she entered Washington University in St. Louis and received her B.A. degree in Government in 1966. She was a legislative assistant and press secretary to Representative Richard Ichord of the Missouri from 1969 to 1973; and then press secretary and legislative assistant to Representative James Symington, also of the Missouri. That same year she became press secretary to Senator Bob Packwood of Oregon. She was promoted successively to administrative assistant, chief of staff, and legislative director for Senator Packwood for the years 1975-1976.

She married Rhett Brewer Dawson on January 15 of that year and retired to raise a child, Elizabeth Stewart. In 1981 Dawson was appointed to the Federal Communications Commission by President Ronald Reagan. Dawson early developed a reputation in the FCC as "one concerned about maintaining the integrity of the spectrum". On November 30, 1981 at a black lawyer's conference, Dawson said minorities should concentrate more on the opportunities presented by new technologies to increase their presence in telecommunications. Dawson felt if minorities were to play an important role in ownership, they'd have to focus on acquiring existing communications properties to enter new markets.

Dawson noted that since 1978, when the FCC adopted its policy statement on minority ownership, there had been an increase of about 600 new broadcast stations, but an addition of only about 65 minority-owned broadcast outlets. Commissioner Dawson also noted that if such a trend continued over the next 10 years, minorities would own only 5 percent of all broadcast properties.

On the other hand Dawson said the expansion of the tele-communications market, the Commission's new entry policies and the general revision of the Commission's licensing policies, would create the greatest opportunities for minority participation. Dawson believed that small businesses and minorities would be presented with "real opportunities" and "enhanced services" which combined data processing applications with common carrier functions. Other areas Dawson said could provide minorities with ownership opportunities were low-power television (LPT) and Direct Broadcast Satellite (DBS) systems. Regarding the licensing process, Dawson felt using a lottery to award licenses could work to the benefit of applicants who might have insufficient resources to make it through comparative procedures. Dawson also believed economic qualifications should not pose a barrier to getting a license. She thought that FCC trafficking rules "should be slanted for a quick exit." Commissioner Dawson said some of those rules were incompatible with a free market and did very little to serve the goals of promotion of minority ownership in broadcasting.

Dawson promised to see that more minorities were placed in decision-making positions at the FCC. She demanded assurance that proposals for new services would not result in erosion of existing ones. She said the Commission "must be aware of the point at which the effectiveness of the broadcast medium as a competitor of the other media is reduced." Commissioner Dawson said she was as pro-competition as any member of the FCC. She left the FCC on December 4, 1987.

By Gerald V. Flannery, Ph.D. and James Nunez, M.S. candidate.

PATRICK, DENNIS R.
1983-1992

Described as tough-minded and independent, Dennis Patrick joined the Federal Communications Commission during a time when deregulation was the rule. Patrick, a republican, was nominated by President Ronald Reagan and sworn in on December 2, 1983. He joined a Commission that had been under fire for its lack of independent thinkers because the Commissioners would rarely cast dissenting votes against Chairman Fowler on key issues.

Born in Glendale, California on June 1, 1951, Patrick pursued a legal career prior to his nomination to the FCC. He received an A.B. degree from Occidental College in 1973 and graduated *magna cum laude*. His hard work earned him a spot in Phi Beta Kappa. Patrick went on to the University of California at Los Angeles, (UCLA) where he earned his J.D. degree in 1976.

During law school, Patrick served as a law clerk to the Honorable Justice William P. CLark of the California Supreme Court. After graduating, he practiced law with the Los Angeles firm of Adams, Duge, and Hazeltine from 1976 to 1981.

Patrick then took his first government position, one that would eventually lead him to the FCC. He joined the White House as the Associate Director of Presidential Personnel. Patrick was responsible for dealing with legal and regulatory agencies including the FCC, and he recruited and interviewed candidates for presidential appointments. He held the position from 1981-1983 until he moved into another government position. In 1983, Patrick served as the Special Assistant to Administrator of the National Telecommunications and Information Administration of the Department of Commerce. That position set the stage for his membership in the FCC. Patrick, however, had to stand by idly for months before his confirmation as Senator Barry Goldwater and the bureaucratic system combined to slow the process. Six months of Patrick's tenure passed with the FCC before the Senate made it legal. Patrick had been granted a recess appointment. The Senate Commerce

Committee voted on April 2 to permit Patrick to fill the remainder of the term vacated by former Commissioner Anne Jones. Goldwater later apologized for his interference.

During his first term, the FCC voted 3-1, to require commercial television broadcasters to serve the special needs of the children and the community. In addition, it ruled that licensees would have broad discretion in determining how best to see that children and the community are properly represented and served.

The move toward deregulation played and still plays a commanding role in the Commission's rulings. Patrick's first deregulatory exposure was to the FCC 4-1 vote to repeal and / or relax the regional concentration rule.

Patrick wasted little time expressing his views on deregulation. In his first public appearance, Patrick told the National Cable Television Association that he felt the marketplace was the best regulator. The following month, in July 1984, the FCC voted unanimously to deregulate television. The decision allowed television licensees to make decisions on program length, commercial and non-entertainment programming. The licensees would still, however, be obliged to provide programming that is responsive to the issues. The same decision extended relief to non-commercial radio and television broadcasters, eliminating their ascertainment and program logging requirements. In another important ruling, Patrick was involved in the FCC vote to eliminate the 7-7-7 ownership rule and to incorporate a 12-12-12 ruling that would be retroactive as of 1990.

Patrick received reappointment to the FCC which was confirmed by the Senate on July 19, 1985. He was sworn in on July 25 to a term that was scheduled to end June 30, 1992.

In March of 1986, Patrick and the FCC voted unanimously to eliminate all restrictions on duplication of programming by AM-FM combinations. The deregulatory move was expected to expand the hours of operation and promote improved service to the public. Then, Patrick was part of a February 1987 FCC unanimous vote to reinstate syndicated exclusivity rules and also began proceedings to build a case against cable television's compulsory copyright license. On April 6, 1987, FCC Chairman Fowler announced that he would step down on April 17th and would be replaced by Dennis Patrick. On April 18, 1987, Patrick assumed the Chairmanship of the FCC.

In his first comments as Chairman, Patrick said he hoped his words and deeds would mark him as a responsible deregulator who would approach every issue from the perspective what is most consistent with the law and the public interest".

Patrick consistently showed independent thought and thorough research. He was a careful administrator and an independent thinker. Broadcasting magazine said: "It is Chairman Patrick's chief objective to see that the publics' interest is served and that Congress will be supportive of his efforts to 'maximize' the public interest."

By Gerald V. Flannery, Ph.D. and David A. Male, M.S.

DENNIS, PATRICIA D.
1986-1989

Federal Communications Commissioner, Patricia Diaz Dennis was born in Santa Rita, New Mexico on October 2, 1946. She is the daughter of Portfinio Madrid and Mary Romero Diaz. On August 3, 1968, she married Michael John Dennis. They had three children; Ashley Elizabeth, Goeffrey Diaz, and Alishia Sarah Diaz and lived in northern Virginia.She was the second Hispanic on the FCC, a Democrat and a Roman Catholic as well. She received an A. B. in English from UCLA in 1970 and in 1973, she received her J.D. from Loyola University, Los Angeles. She was on the dean's list at both Universities.

Dennis passed the California Bar in 1973. She was a law clerk and a California rural legal assistant in 1971. From 1973-1976, she was an law associate with Paul, Hastings, Janofsky and Walker in Los Angeles. She was the attorney for the Pacific Lighting Corporation in Los Angeles from 1976-1978. She later became the Assistant General Attorney for the American Broadcasting Company in Hollywood, California (1978-83). From 1983-1986, Dennis devoted her efforts toward her membership with the National Labor Relations Board in Washington, D.C. On June 13 she was confirmed by the Senate as the fifth member of the Federal Communications Commission. A brief list of her honors would include:

Executive Editor of the Loyola Law Review 1972-27
Board of Directors for the National Network of Hispanic Women 1983-84
Committee member of the Coro Foundation Hispanic Leadership Program 1981-82
Board of Directors of Resources for Infant Educarers 1981-83
Recipient of Certificate of Achievement for YMCA 1979
Member of the Mexican American Bar Association
Los Angeles County Bar Association: Child Abuse Subcommittee Chairman
Hispanic Bar Association, D.C.
Received the Fouragere Honors at graduation from UCLA

Member of the Bar Associations of the Supreme Court of the U.S.
Member District of Columbia Court of Appeals
Recipient of the Mexican American Opportunity Foundation Woman of the Year Award
Member of the United States Delegation, United Nations Decade for Women, 1985 World Conference-- Nairobi, Kenya.

When The New York Times reported the appointment of Dennis to fill the vacancy on the FCC, she had virtually no experience with communications. She was a lawyer with only minimal government experience who had been working for the National Labor Relations Board (NLRB) for three years, and who had authored articles on labor law. She did, however, mingle in the world of communications during her two year stay as Assistant General Attorney for ABC.The lack of experience apparently did not bother President Ronald Reagan, who appointed her. It did not bother Congress who confirmed her, nor did it bother Dennis, who discussed her lack of experience candidly in her opening statement at the hearing to decide her confirmation. Several weeks before her confirmation, (June 13, 1986) Dennis was reported as "boning up on technical aspects and on matters pending" (The New York Times, May 26, 1986). Also at this time, the House approved a Senate passed bill reducing the terms of FCC members to five years from a previous seven years. Because Dennis was a Democrat, her appointment broke a decision making grid-lock that existed in the agency.

The FCC at this time was under the chairmanship of Mark Fowler, one of the most controversial chairmen and the longest serving Chairman in FCC history. He was famous for his views on the deregulation of the broadcasting industry. He and President Reagan were able to abolish the Fairness Doctrine, the law that requires broadcasters to present contrasting viewpoints on controversial issues.

Dennis breezed through her confirmation hearing. She was not asked to respond to any substantive questions. She merely made a three minute opening statement and after a short recess, Senator Barry Goldwater told her she was "in." She attended her first meeting of the FCC on June 26, 1986, but she did not vote on any items on the agenda. She was believed to have no allegiance to any member of the Commission, not even Mark Fowler who needed her support to give the swing vote on major issues.

In her first vote as a Commissioner, Dennis gave Fowler and Dennis Patrick, fellow FCC Commissioner, the third vote they needed to overcome opposition by Commissioners Quello and Dawson in a controversial spectrum allocation question. Several weeks later, she voted on the major issue concerning the 1986 cable rules of must-carry, the requirement that cable systems carry all local broadcast signals meeting some arbitrary criteria. The arbitrary criteria being, the requirement that cable systems see to it that A / B switches were installed in the homes of cable subscribers so that they could receive all local broadcast signals off the air with the flip of a switch. It was a five year rule obliging cable operators to make A / B switches available to subscribers (that rule later came under reconsideration). Dennis remained an FCC Commissioner after Fowler left.

The FCC's newest female (then) and Hispanic Commissioner was sworn in by Vice-President George Bush and was the sixth woman in history to hold a seat on the Federal Communications Commission.

By Gerald V. Flannery, Ph.D. and Carla P. Coffman, M.S.

SIKES, ALFRED
1989-1993

Alfred Sikes was nominated to the Federal Communications Commission by President George Bush and confirmed by the Senate on August 4, 1989. He served as FCC Chairman until 1993. He had a long history in broadcasting and politics, a nice combination for an FCC Commissioner. He was formerly Assistant Secretary of Commerce and Administrator for of the National Telecommunications and Information Administration at the Department of Commerce. Prior to that he headed his own company Sikes and Associates, Inc. a broadcast management and media consulting firm. He also served as an officer of a number of companies that owned and operated radio stations in Texas, Louisiana, and New Mexico. Sikes had the experience necessary to understand broadcast regulations.

Sikes was born on December 16, 1939 in Sikeston, Missouri. He received his education from Missouri schools, with a B.A. degree in 1961 from Westminster College at Fulton and a law degree from the University of Missouri Law School in Columbia in 1964. After school he joined the firm of Allen, Woolsey and Fisher.

His broadcast background also included the ownership and the principal ownership of five Missouri radio stations. He served as Director of the Missouri Department of Consumer Affairs, Regulation and Licensing, Director of the Missouri Transition Government of Governor-Elect Christopher Bond, whose campaign he had earlier managed, and also as Assistant Attorney General for the State of Missouri. Work like that led to a close working relationship with John Danforth (R-MO), a member of the Senate Commerce Committee. Sikes ran his campaigns in the 1970's In return, Danforth supported Sikes in his political and broadcasting endeavors.

Sikes had become professionally known as a hardliner on issues that arose while he held the Chairmanship as he worked to implement the agency's rules and regulations. Not long after being named to the position, Sikes voted to deny license renewal for two broadcast stations. He told Broadcasting magazine that he thought regulation was of utmost importance and that broadcasters acted as "public trustees." To him that mean that every station would ultimately be responsible for mis-representation made to the FCC or even to such practices as the double billing of advertisers. This "tough-nosed" style of Sikes was thought to be an acquired manner growing out of a combination of his political and broadcast history. At any rate, he put teeth into some of the FCC regulations that broadcasters had generally ignored.

Public interest seemed to be foremost on his mind and to figure consistently in his decisions. In what has been called the FCC's most severe action ever, New York Telephone and New England Telephone & Telegraph were ordered to pay a $1.4 million fine and to refund $35.5 million to their ratepayers in the NYNEX case. The penalties were a result of high prices passed on to ratepayers; and the case was a prime example of Sikes' intolerance with public deception.

The NYNEX case is only one example of how the FCC, under Sikes, was quick to identify wrongdoing and punish it. A consent decree demanding a contribution of $1 million from Centel Cellular Telephone Company to the United States Treasury was signed by the FCC. Along with this steep fine, the company also had to conduct an educational program on the subject of radio tower markings and lighting, after a accident resulted in the deaths of two people whose helicopter collided with a Centel tower. Later, there was a near helicopter collision with another one of their towers.

Broadcasting magazine on August 13, 1990, quoted Sikes as saying, "The Commission needs to act forthrightly...we shouldn't just let cases sit here. And so you can expect, from this Commissioner, that kind of forthright action."

That is exactly what Sikes did. After only five months as FCC Chairman, he voted to deny the renewal of KQET and WBBY's licenses. Former FCC Mark Chairmen Fowler and Dennis Patrick had let those cases sit for three years without any action. Sikes put his verbal promises into action. At the agency he promoted high-definition television (HDTV) and wanted to see the phone companies be able to get into the video dial-tone business and provide movies and TV programs over the phone. On May 29, 1990 on the "Larry King Show", Sikes said his stand on enforcing the FCC's anti-indecency standard, that regulated when certain kind of material could be shown, could be viewed as censorship, but he felt that children must be protected regardless of the consequences. Censorship might have to be that sacrifice. He received strong support from many senators for his stance against indecent programming. The senators said there must be some balance struck in that area and they believed Sikes had keyed into that perception. Sikes' believed there must be a balance between First Amendment protection of broadcasters and the protection of children from indecent material. On October 9, 1989 Broadcasting magazine quoted Sikes as saying, "I believe that with freedom comes responsibilities and that the enemy of freedom is irresponsibility." He went so far as to support a 24-hour indecency ban to prohibit that type of programming.

Sikes, told a meeting of the International Radio and Television Society:

> I can remember once-thriving motels, that, when passed by an interstate highway,chose to compensate by installing waterbeds, VCR's with X-rated flicks, and rental by the hour. Their survival was fleeting. Commercial success in broadcasting will not be ensured by mindless shows built around sex, gratuitous violence, or the latest effort to exploit bizarre conduct, advanced by so-called talent researchers.

Sikes referred to himself as the "adaptive regulator" saying regulation was only imposed, after an evaluation of the market, resulted in little or no competition. Another controversial thing he strongly supported was a proposal to give existing television stations another channel, without having to compete for it, to program it whatever way they wished, whether it be movies, entertainment, sports, business or a shopping service.

Did this former Missouri prosecutor sometimes make bold statements prematurely? Possibly yes. In 1989, the White House delayed the announcement of his nomination to the FCC Chairmanship after Sikes leaked that information to The New York Times before the official nomination. However, despite his eagerness with some issues, Sikes took a forthright stance on issues of public concern and acted openly.

Sikes left the FCC on January 19, 1993, and at 53 years of age, later accepted a position with the Hearst Corporation heading a new operating group to oversee Hearst's interests in new technologies. He told Electronic Media magazine that Hearst would be looking for new opportunities in electronic media brought about by the interaction of the computer, telephone and high powered satellites.

By Gerald V. Flannery, Ph.D.

MARSHALL, SHERRIE
1989-1993

Federal Communications Commissioner Sherrie Marshall was appointed to the FCC by President George Bush in the summer of 1989. The Senate confirmed the nomination on August 4, and she was sworn in two weeks later. Sherrie Marshall was born on August 3, 1953 in Jacksonville, Florida, the oldest of two children to insurance executive Clifford Marshall. When she was eight years old, her family moved to Durham, North Carolina where she set her sights on becoming an astronaut or a cowboy. At 13 years of age, she was selected to serve as a page in the North Carolina Senate and as a result, she reset her sights on law and governmental service.

In 1974, only three years after she started college, Marshall graduated from Chapel Hill with her Bachelors degree in English. She immediately started law school, and, two years later, served on the advance team for Republican vice presidential candidate Robert Dole. As a member of the advance team, she selected sites for Dole's appearances, made sure the logistics were what they wanted, and insured an enthusiastic welcome from local media personalities and dignitaries. Marshall told Broadcasting magazine her work with the advance team made her learn to pay attention to detail and become better organized. One year later, in 1977, she received her law degree from Chapel Hill.

Marshall did not waste any time getting into law and government; upon graduation, she landed a position with the Federal Election Commission (FEC), as an attorney in the office of the FEC General Counsel. There, she assisted with the implementation of new election laws.

In 1978, she married attorney Paul Cooksey who she met on the Hill in the late 1970s. During this time, she also accepted a position with the Senate Committee on Rules and Administration where she offered advice on election laws to Republican Senators. One year later, FEC Chairman Max Friedersdorf recruited her to serve as his executive assistant. In 1981, Friedersdorf was appointed head of legislative affairs for the Reagan White House. Marshall went along as Special Assistant to the President for Legislative Affairs, where she counseled such clients as Supreme Court Justice Sandra Day O'Connor and FCC Chairman Mark Fowler on the Senate confirmation process of nominees. In 1982, she was offered a position as a staff member under Chief White House Counsel Fred Fielding where she focused on ethics and election laws.

Marshall first met President George Bush when she was asked to help prepare him for the 1984 debate against Democratic vice presidential candidate Geraldine Ferraro. Her preparation for the Philadelphia debate proved effective, and Bush contacted her immediately after the debate to thank her for her help. When White House Chief of Staff James Baker was appointed Secretary of the Treasury in 1985, Marshall became his executive secretary. She remained in that capacity until the law firm of Ropes & Gray recruited her in 1986. Within months, she left Ropes & Gray to become a partner in the Washington D.C. law firm of Wiley, Rein, & Fielding.

FCC Chairman Dennis Patrick opened the door to the Commission for Marshall in 1987 when he offered her the position of Director of Congressional Affairs. During this time, the FCC was trying to smooth things over with Congress after it had repealed the Fairness Doctrine. As a member of Patrick's staff, she began attracting a large audience in the communications community. Unfortunately, the audience was not fully supportive of her efforts at improving Patrick's relations with the legislators. As if that point in her life weren't chaotic enough, she and her husband of eight years separated and divorced.

When Bush was elected president in 1988, Marshall was appointed senior member of the president's legal staff. In that capacity, she offered ethical and related advice in bringing Cabinet officers and other presidential appointees through the Senate confirmation process. After the transitional period lapsed, she returned to Wiley, Rein, & Fielding until her appointment to the FCC.

Marshall took the time to dismiss rumors that her position with the FCC would enhance the influence of former law partner Dick Wiley. She went on record as saying Wiley would be influential whether or not he talks to her. She said if she discussed an issue with him, she'd discuss it with all of the other parties involved as well.

Commissioner Marshall brought a conservative, Republican faith in competition to the FCC along with a high regard for feasibility, fairness, and public benefits. She had few qualms about decency and indecency and believed broadcasters should make their own informed decisions regarding those issues. She argued, however, that broadcasters should also consider the guidelines given to them by the FCC before airing such material.

Marshall said her basic goal was to understand the "practical ramifications of what our policy decisions will do." To that end, she announced, that after the Commission, she was forming her own consulting company in Los Angeles and that her first client was one of the better known entertainment law firms in Hollywood. During her time on the Commission, Marshall was known as one of the strongest supporters of Hollywood, particularly in the area known as fin-syn rules. Marshall said she picked Hollywood since she thought that's where the industry's future was brightest.

By Gerald V. Flannery, Ph.D. and Kim Bourque, M.S.

BARRETT, ANDREW C.
1989-

Andrew Camp Barrett, the third black commissioner in the history of the Federal Communication Commission and the first black to serve on the FCC in nearly a decade, brought a strong background and moral commitment to the agency.

Barrett was born in Rome, Georgia on April 14, 1940 but grew up in Chicago in a Irish-Catholic neighborhood where his father owned the Silver Rail, a "shot-and-a-beer" tavern. After graduating from a parochial high school, he worked for the postal service and at other odd jobs while attending the University of Chicago part-time. He served as a sergeant in the U.S. Army from 1963-65 then went on to work as unit manager of Chicago's Metropolitan YMCA from 1966-68. In 1969, he earned a political science degree, one of his many academic credentials, from Roosevelt University of Chicago. He was active in the civil rights movement while collecting other scholarly credentials, which include a Masters in Public Administration and Economics from Loyola University in Chicago and a law degree from DePaul University. During this movement for human rights Barrett, then a Democrat, was the associate director of the National Conference of Christians and Jews from 1968-1971 then became the executive director of the National Association for the Advancement of Colored People (NAACP) until 1975.

Barrett's record revealed he was a fighter for human rights. While working with the National Conference of Christians and Jews he was involved in managing the group's educational and human relations programs. While most activists picketed and demonstrated on the streets in the 1960's, Barrett and other NAACP members went to federal court to protect the rights of women and minorities.

After receiving his law degree, Barrett worked for Jimmy Carter in 1976 and shortly after, was chosen by then Democratic Governor Dan Walker to be director of operations for the Illinois Law Enforcement Commission. At that agency, where federal and state funds were disbursed to local courts and police, Barrett was responsible for agency spending and disbursement to the approximately 450 recipients of the grants.   James Zagel, Barrett's boss at the agency, who became a federal district court judge in Chicago, discussed Barrett in the August 1989 issue of Broadcasting magazine.   Barrett was described as a man who was not familiar with many of the FCC issues but had the credentials necessary to deal with the congressional politics that were thought to be at the center of some FCC decisions. Zagel said Barrett had held many difficult jobs that required plenty of judgment and had "a gift for giving people bad news without driving them up the wall.".

Shortly after the 1976 Chicago gubernatorial election, when a republican won over a democrat, Barrett switched to the Republican party. The August 1989 issue of Broadcasting magazine reported he did not switch to gain popularity with the newly elected Republican Governor Thompson, rather he switched because it gave him more opportunity and he felt the local Democratic party began taking the black vote for granted.

In 1979, Governor Thompson chose Barrett to play a key role in the combination of three state agencies, involved in economic development, as the Illinois Department of Commerce and Community Affairs. As a result, the Illinois Department of Commerce was created with Barrett as assistant director. In 1980, Thompson gave Barrett a seat on the Illinois Commerce Commission, which was a seven-member board responsible for regulating electric, gas, telephone and water in Illinois, a state referred to in Black Enterprise magazine as "one of the nation's most deregulatory minded."   It was on this board that Barrett gained 9 years of experience with regulatory commissions preparing him for his FCC position and earn him the title of "master of the bureaucratic art", as noted in the October 1991 issue of Ebony.

Barrett hoped for an FCC appointment in 1988 but President Ronald Reagan nominated Washington attorney Susan Wing and FCC staffer Bradley Holmes. Those nominations were never confirmed by Congress, and then FCC Chairman Dennis Patrick resigned in 1989. When George Bush became President, three seats on the FCC were vacant. Barrett served the last eight months of another Commissioner's term and was renominated by President Bush for a five-year term.

Barrett was confirmed in August 1989 at a tense Senate confirmation hearing that concentrated on the issues of indecency and violence in programming and the recent problems of the FCC. Jet magazine for June 18, 1990 said Barrett was praised by Senator Robert Dole (R-Kans) for "eight months of demonstrated ability" but he later raised eyebrows and even angered some of the senators with his unsettling responses to several questions. When questioned by committee members on indecent and violent programming, Barrett said he felt both were offensive but indecent programming would not be a part of broadcasting if there were not a market for it, so he would have to consider the market demand. Barrett's answer angered several senators who were still upset with former FCC actions allowing the marketplace to guide Federal Communications policy. To clarify his comments on the indecency issue, Broadcasting magazine, in August 1989, reported Barrett later wrote to Senate Commerce Committee Chairman Ernest Hollings (D-S.C.) stating "there is no question that preventing obscene and indecent broadcasts is a critical function the FCC." Barrett also earnestly agreed to do "everything in my power under the law to ensure that licensees across the nation broadcast programs that comport with the highest moral standards of our society" Barrett later reinforced his views on indecency by saying, as quoted in the February 1990 issue of Black Enterprise magazine,

> I think we Commissioners have an obligation to enforce the rules on the books. Indecency is an area where we must weigh First Amendment concerns versus our statutory obligation to prevent the airing of obscene, indecent programming.

Barrett began his work for the FCC in September and his voting record showed him to be a deregulatory pragmatist, ready to strike regulations but not until he weighed the practical impact that action would have on the affected businesses. He and fellow Commissioners Duggan and Marshall, formed a three-vote majority that resulted in changes in proposals introduced by FCC Chairman Alfred Sikes.

Barrett's appointment came at a time when there were conflicts between Congress and the FCC. Congress was not satisfied with the strict conservative ideology of the FCC before Barrett and his fellow Commissioners arrived. Channels magazine, writing in September 1990, reported the FCC was comprised of "independent thinkers and strong-willed personalities." Many said the new variety of personalities and different opinions was a positive and much needed change for the agency and the relations between FCC and Capitol Hill had improved tremendously. The FCC demonstrated some of that cooperative spirit in discussing their first issue , the cable report, when they reached a consensus with a 5-0 vote to approve it.

Barrett played a key role, an unexpected one to many, in the FCC's second issue of the battle of syndication regulations of television rerun rights and revenues. He devised the plan that changed the 21-year-old FCC financial interest and TV rerun "fin-syn" regulations. His compromise restricted network ownership to 40 percent of their prime-time shows while giving networks more opportunities to sell reruns in foreign markets. Those rules limited the power of television's three largest networks, but gave unknown writers and producers an opportunity to sell their work in the big league. The October 1991 issue of Ebony magazine summarized Barrett's response:

> I don't fool myself to think there is going to be an enormous amount of jobs created for poor people or people who are underemployed or unemployed...But the well being of those minor writers and producers who would not have survived before the "fin-syn" modification depends on those rules staying in place.

The Barrett "fin-syn" proposal had been controversial with some praise and some complaints arising. The new rules went to the U.S. Court of Appeals in Chicago which considered petitions for review from those who said the new rules were too restrictive and those who felt they were too lenient. Barrett said the plan was equitable and it would have been unfair for the networks / major studios to walk away with everything.

Barrett also showed concern for women and minority ownership of mass media properties. The fact that minority ownership policies exist, he said, acknowledges the under-representation or misrepresentation of women and minorities in the broadcast media. He said women and minorities had made significant gains in the ownership of broadcast properties but the number was sparse when compared to the 11,000 existing broadcast stations.

By Andrea LeBlanc, M.S. and Gerald V. Flannery, Ph.D.

DUGGAN, ERVIN S.
1990-1994

Ervin S. Duggan was born in Atlanta, Georgia on June 30, 1939 just as war was about to spread in Europe and the headlines were filled with stories about Germany and the use of propaganda techniques by Hitler and Goebels. America was coming out of a serious depression, the likes of which it had never seen, when his family moved to Nanning, South Carolina. President Franklin D. Roosevelt was in office, serving his third term, and radio was firmly entrenched as a major entertainment medium in the United States.

Duggan's early years spanned World War II, the post war boom and the introduction of television which began to sweep America in the late 1940's. He was a young boy when Milton Berle and Ed Sullivan and Edward R. Murrow were household words: he grew up with television. After high school, he went to Davidson College in North Carolina and graduated with a B.A. in 1961; then he became an officer in the U.S. Army and served from 1962 to 1964. After his discharge, he went to work for the Washington Post as a reporter and stayed there for two years.

From 1965 to 1969 he worked as a Staff Assistant for President Lyndon Johnson, then moved over to the Smithsonian Institution as Director of Special Projects, History and Art. He turned his hand again to writing in 1970-1971 and became an author with Doubleday and Company. Next he teamed up with U.S. Senator Adlai E. Stevenson working as a Special Assistant from 1971-1977. In 1979, he joined the U.S.Department of State as a member of the Policy Planning Staff and stayed there for two years, then went on to become the National Editor of the <u>Washingtonian</u> Magazine from 1981 to 1986. In 1981 he also founded Ervin S. Duggan Associates, his own Washington-based communications consulting firm. On February 15, 1990, President George Bush named him to take the democratic seat on the Federal Communications Commission; he was confirmed by the Senate on February 15, 1990 and sworn in two weeks later

<u>Broadcasting</u> magazine reported that Duggan, introduced and supported by Democratic Senator Ernest Hollings, Chairman of the

powerful Commerce Committee, and supported by South Carolina's other Senator Strom Thurmond, sailed through his confirmation hearings and was viewed by the committee as a person who believed in "family values," a term in coinage at that time, someone who would rid the airwaves of "indecent material" and bring broadcasting back to the high moral ground. The indecency problem centered around something dubbed the "Safe Harbor" concept of programming, meaning that there should be some part of the day when adult material could be broadcast without violating the FCC indecency standard for licensees.

Cable operators and channels were able to transmit movies that had problem content at any time of the day or night since they were involved in "point to point" communication, which resembles a private phone call, that is between two parties. Broadcasters, however, sent their signals into the open air and their programming could be picked up by anyone who had a set / antennae / rabbit ears, thus adult material was often relegated to the late night hours. To allow broadcasters access to the kinds of material (movies, comedy specials, mini-series, etc.) that may have had some objectionable sections, but was available to cable operators, the FCC tried creating a time period (Safe Harbor), first from 10 p.m. till 6 a.m., then modified it to from midnight to 6 a.m. It continued to stir controversy.

Duggan told supporters he had great respect for the First Amendment and would not "trash" it, referring to his years as a journalist, saying it had great meaning for him. The 50-year old Duggan assured Senator Hollings that he believed in a light hand with regulation but also felt that some regulation was necessary. For example, he said his top priority was to redefine the "public interest" standard in light of the present-day communications industry, to try to get broadcasters to voluntarily return to some of the standards they were no longer required to follow by law: areas like news and public affairs, the concept advanced by the repealed Fairness Doctrine, children's programming, involvement with community causes, and minority rights in programming and employment.

Broadcasting magazine, in its latest chronology of important events , listed a number of things the FCC was involved in during the early nineties. Some of them were: the cable industry promised to change its customer service policies with adoption by National Cable Television of a group of comprehensive standards; the U.S. government service aimed at Cuba, TV Marti, went on the air, only to be heavily jammed by Cuba; the FCC came forward with a package of rule changes designed to help AM radio; Chairman Alfred Sikes ordered an outside review of whether the broadcasting marketplace could better regulate the industry than the FCC; approved a plan to limit commercial time on children's programs; and President Bush signed a bill requiring television sets over 13 inches to display closed caption material.

Duggan left the FCC in 1994 to head the Public Broadcasting Service.

By Gerald V. Flannery, Ph.D.

HUNDT, REED E.
1993-

President William Clinton, facing several vacancies at the agency, named 45-year old Reed Hundt, a democrat, as Chairman of the Federal Communications Commission, saying he felt sure he would do a good job steering the FCC over the coming years of growth and change. Hundt was confirmed by the Senate on November 19, 1993 and was sworn in on November 29th.

Hundt, born in Ann Arbor, Michigan on March 3, 1948, received a B.A. in History, *magna cum laude*, with exceptional distinction, from Yale College in 1969 and a J.D. from the Yale Law School in 1974. He married Elizabeth Ann Katz, a psychologist and they had three children. Hundt served as law clerk to the late Chief Judge Harrison L. Winter of the U.S. Court of Appeals for the Fourth Circuit. He was a member of the District of Columbia, Maryland and California Bars.

Hundt was a senior partner at Latham and Watkins, a national and international law firm, before coming to the Commission, where he was involved in state, federal and international communication issues, including things like local exchange telephone service, long distance, international fiber optics, cable television alternatives, satellites, and First Amendment work. He was primarily an antitrust lawyer who worked on several cases involving telecommunications companies, Bell Atlantic and Turner Broadcasting to name two. He also did pro bono work for the United States Court of Appeals for the Fourth Circuit, the NAACP Legal Defense Fund, the Lawyers Committee on Civil Rights, Conservation International and the D.C. Preservation League.

Hundt, a classmate of President Clinton at Yale, served as senior advisor to the Clinton for President campaign and as a senior advisor and member of the Economic council for the Presidential Transition Team. In addition, he had advised Vice President Albert Gore, whom he knew since their days at prep school, on economic issues since 1984. Hundt told reporters early in is tenure that the Clinton administration was going to review the rules and policies that kept broadcasters from entering other new telecommunications businesses. That news came like a breath of fresh air to station owners, along with Hundt's declaration that the FCC was going to take another look at its ownership rules in view or more deregulation. But his presence drew both criticism and praise.

Bills introduced in Congress, at that time, were designed to allow local telephone companies, long distance services, cable operators and other telecommunications businesses to enter and compete with each other. The telecommunications sector was expected to grow to one trillion dollars in revenue by 1996, and Hundt felt it would represent a sixth of the national economy by the year 2000. Hundt thought that the decisions the FCC made in those areas could result in stimulating the nation's economy and that competition among communications entities was in the" public interest" because it would help insure that America's information superhighway brought inexpensive services to homes.

The FCC operated through 1993 with only three Commissioners instead of five. Industry leaders saw the major problem facing the FCC, in 1994, as one involving the White House and Congress, grappling for the upper hand in naming new Commission members, and the subsequent delays in naming those members, since the Commission was supposed to be fairly balanced between Democrats and Republicans, and was supposed to represent a diversity of viewpoints, political and otherwise. Critics claimed that the delay in naming new members, amounted to letting the U.S. communications infrastructure be overhauled in the marketplace, directed by the biggest and most powerful players.

An unannounced meeting between Hundt and Vice President Gore, which took place in the spring of 1994, a few days before the FCC's controversial ruling on cable rate cuts, (more were required), set tongues wagging about who was running the agency. However, insiders quickly assured reporters that at least 12 people were at that meeting, and that it dealt with helping Gore plan for an upcoming tele-communications conference in Argentina. Later, Hundt passed up a chance to address / meet with some of the 70,000 people interested in / connected with broadcasting who attended the April 1994 NAB meeting in Las Vegas. He chose to attend that Buenos Aires conference with Gore. Critics again felt that Hundt was aligning himself more with public interest groups and the White House, and that the entire FCC was tilting more toward a consumer-oriented policy.

Two examples of this were said to be the FCC's decision to again control cable rates by rolling prices back to their levels as of September 30, 1992 and its broad-caster decision, the same day, on financial interest and network syndication rules.

By Gerald V. Flannery, Ph.D.

CHONG, RACHELLE
1994-

President William Clinton openly discussed his intention to name more women to the Federal Communications Commission but the vacant seats remained empty until two women went before the Senate committee for confirmation in May 1994. They were Rachelle Chong, a communications industry lawyer, and Susan Ness, a Washington attorney, women who were expected to be approved, and who would be the eight and ninth females to serve on the FCC.

Rachelle Chong, a 35 year old woman of Asian extraction, was then a law partner in the San Francisco firm of Graham and Jones, specializing in telecommunications law, and also working as a litigator and adviser on public utilities issues. She was expected to fill the Commission's Republican seat. Ms. Jones numbered among her clients McCaw Communications, said to be the nation's largest cellular radio company, a company being acquired, at that time, by American Telephone and Telegraph (AT&T), and Pacific Telesis, one of the former "Baby Bells" created by the breakup of AT&T and its regional phone operations. She also practiced before the FCC from 1984 to 1987.

Senate Minority Leader Robert Dole (R-Kan.) supported her nomination. The Washington Post wrote that lobbyist Ronald F. Stowe, who represents Pacific Telesis, said: "She's a very bright, very pragmatic and independent person who knows a lot about telecommunications issues. I've been very impressed with her grasp of issues."

Reed Hundt, then Chairman of the FCC, told an industry gathering in February 1994 that he thought they'd be "impressed by these brilliant women," and communication lawyers, regulators and lobbyists seemed to agree. The telecommunications sector was expected to grow to one trillion dollars in revenue by 1996, and Commissioner Hundt felt it would represent a sixth of the national economy by the year 2000.

A number of problems faced the new Commissioner. They were: how to resolve the entry of the telephone companies into the cable and video business, and how to deal with cable companies getting into the phone and data transfer business, personal communications systems, and advanced data networks. This area was earlier known as the Telco problem but had expanded beyond that as days passed. There were numerous other areas they would consider: the mushrooming number of channels, 500 predicted by 1995, 1,000 by the year 2000; the FCC's indecency rule and the Safe Harbor controversy; the probability that the Fairness Doctrine will be reintroduced in Congress and probably supported by President Clinton; the technological ability that through digital compression ten new broadcast channels can be compressed into the place where only one existed before; whether Public Access Television would still be viable when so many channels existed offering every conceivable variety of program; the problem of the growing concentration of broadcast / cable / satellite properties and the conglomerates that created them, plus the increasing pressure for multiple and cross ownership mergers; and finally how the FCC could affect programming content without controlling it.

FCC Chairman Hundt said the Clinton administration was going to review the rules and policies that kept broadcasters from entering other new telecommunications businesses, a proposal that could engage the Commission fully for years.

By Gerald V. Flannery, Ph.D.

NESS, SUSAN
1994-

President William Clinton said he intended to name more women to the Federal Communications Commission but the vacant seats remained empty until two women went before the Senate committee for confirmation in May 1994. They were Rachelle Chong, a communications' industry lawyer, and Susan Ness, a Washington-area investment banker, women who were expected to be approved, and women who would be the eight and ninth females to serve on the FCC.

Susan Ness, 45 years of age, who specialized in communications companies, was a senior lender and later a group head in the communications-industries division of the American Security Bank of Washington, D.C. Prior to that she was an assistant counsel to the House Banking Committee. The Wall Street Journal reported that she had a masters degree in Business Administration from the Wharton School of Business, and a law degree from Boston College Law School; and she founded and directed the judicial-appointments project for the National Women's Political Caucus. She was said to be personal friend of the first lady, Hillary Rodman Clinton.

Electronic Media magazine reported that Ness, a fund raiser for the Clinton-Gore campaign who lived in a suburb of Washington, D.C. would fill the Democratic opening on the Commission, and that Ness was married to attorney Larry Schneider.

Reed Hundt, then Chairman of the FCC, told an industry gathering in February 1994 that he thought they'd be "impressed by these brilliant women," and communication lawyers, regulators and lobbyists seemed to agree. Bills introduced in Congress, at that time, were designed to allow local telephone companies, long distance services, cable operators and other telecommunications businesses to enter and compete with each other. The telecommunications sector was expected to grow to one trillion dollars in revenue by 1996, and Commissioner Hundt felt it would represent a sixth of the national economy by the year 2000.

A number of problems faced the new Commissioner. They were: how to allow entry of telephone companies into the cable / video business, and how to deal with cable getting into the phone and data transfer business. This area was earlier known as the Telco problem but had expanded beyond that. There were numerous other areas they would consider: the mushrooming number of channels, 500 predicted by 1995, 1,000 by the year 2000; the FCC's indecency rule and the Safe Harbor controversy; the probability that the Fairness Doctrine would be reintroduced in Congress and probably supported by President Clinton; the technological ability that through digital compression ten new broadcast channels could be compressed into the place where only one existed before; whether Public Access Television would still be viable when so many channels existed offering every conceivable variety of program; the problem of the growing concentration of broadcast / cable / satellite properties and the conglomerates that created them, plus the increasing pressure for multiple and cross ownership mergers; and finally how the FCC could affect programming content without controlling it.

FCC Chairman Hundt said the Clinton administration was going to review the rules and policies that kept broadcasters from entering other new telecommunications businesses, a proposal that could engage the Commission fully for years.

By Gerald V. Flannery, Ph.D.

Index of Commissioners

Barrett, Andrew C.	214
Bartley, Robert T.	99
Bellows, Henry A.	4
Brown, Thad H.	7
Brown, Tyrone	181
Bullard, William H. C.	10
Burch, Dean	154
Caldwell, Orestes H.	13
Case, Norman S.	42
Chong, Rachelle	223
Cox, Kenneth A.	132
Coy, Albert W.	90
Craven, Tunis A.M.	51
Cross, John S.	120
Dawson, Mary Ann	200
Dennis, Patricia D.	205
Denny, Charles R.	78
Dillon, Col. John F.	16
Doerfer, John C.	105
Duggan, Ervin S.	218
Durr, Clifford J.	66
Ferris, Charles D.	178
Fly, James L.	60
Fogarty, Joseph R.	184
Ford, Frederick W.	117
Fowler, Mark S.	197
Gary, Hampson	33
Hanley, James H.	19
Hennock, Frieda B.	96
Henry, E. William	129
Hooks, Benjamin L.	163
Houser, Thomas J.	141
Hundt, Reed	221
Hyde, Rosel H.	75
Jett, Ewell K.	69
Johnson, Nicholas	147
Jones, Anne	187

Jones, Robert F.   87
King, Charles H.   123
LaFount, Harold A.   23
Lee, Robert E.   108
Lee. H. Rex   151
Leovinger, Lee   135
McConnaughey, George   111
McNinch, Frank R.   55
Mack, Richard A.   114
Marshall, Sherrie P.   211
Merrill, Eugene H.   102
Minow, Newton N.   126
Ness, Susan   225
Patrick, Dennis R.   202
Payne, George H.   39
Pickard, Sam   25
Porter, Paul A.   72
Prall, Anning S.   48
Quello, James H.   166
Reid, Charlotte T.   157
Rivera, Henry M.   194
Robinson, Glen O.   169
Robinson, Ira E.   28
Saltzman, Gen. C. McK.   31
Sharp, Steven A.   191
Sikes, Alfred C.   208
Sterling, George E.   93
Stewart, Irvin   36
Sykes, Eugene O.   1
Thompson, Frederick I.   58
Wadsworth, James J.   138
Wakefield, Ray C.   63
Walker, Paul Atlee   45
Washburn, Abbott M.   172
Webster, Edward M.   84
Wells, Robert   144
White, Margita E.   175
Wiley, Richard E.   160
Wills, William H.   81